2011 Annual Meeting of the American Electrophoresis Society

(AES)

Topical Conference at the 2011 AIChE Annual Meeting

Minneapolis, Minnesota, USA
16-21 October 2011

ISBN: 978-1-61839-577-1

Printed from e-media with permission by:

Curran Associates, Inc.
57 Morehouse Lane
Red Hook, NY 12571

Some format issues inherent in the e-media version may also appear in this print version.

Printed by Curran Associates, Inc. (2012)

For permission requests, please contact AIChE
at the address below.

AIChE
3 Park Avenue
New York, NY 10016-5991

Phone: (203) 702-7660
Fax: (203) 775-5177

www.aiche.org

Additional copies of this publication are available from:

Curran Associates, Inc.
57 Morehouse Lane
Red Hook, NY 12571 USA
Phone: 845-758-0400
Fax: 845-758-2634
Email: curran@proceedings.com
Web: www.proceedings.com

TABLE OF CONTENTS

(6a) Stochastic Resonance Enhances DNA Transport In the Entropic Trapping Regime

Victor M. Ugaz and Nan Shi, Artie McFerrin Department of Chemical Engineering, Texas A&M University, College Station, TX

Recent interest in entropic trapping (ET) phenomena in nanoporous hydrogels has been excited by their potential to enable enhanced electrophoretic separations of charged biomolecules (e.g., DNA, proteins). For example, an unexpected trend of increasing separation resolution with DNA size has been observed in microchip-based electrophoresis experiments under ET-dominated conditions. But despite this great promise, it has proven challenging to apply these novel effects in practical settings due to lingering problems associated with inherently high diffusion relative to conventional reptation-based transport. Here we show how both the source and solution to this problem are intimately linked to the activated nature of the entropic trapping process. Under a continuous electric field, DNA transport between neighboring pores becomes increasingly uncorrelated over time, leading to undesirable diffusion enhancement. We have found that this condition can be overcome by applying a pulsed electric field at a frequency chosen to match the system's activation events, creating a stochastic resonance effect that can be exploited to significantly improve separation resolution in the entropic trapping regime. The underlying transport principle and effect of decreasing diffusion are demonstrated by our physical model.

See more of this Session: Advances In Electrokinetics and Electrophoresis - Fundamentals
See more of this Group/Topical: Topical 3: 2011 Annual Meeting of the American Electrophoresis Society (AES)

(6b) Accumulation of Air Bubbles In Field Driven Oscillations of a Droplet

Ezinwa Elele, Yueyang Shen and **Boris Khusid**, Chemical, Biological & Pharmaceutical Engineering, New Jersey Institute of Technology, Newark, NJ

Due to the lack of gravity, the removal of bubbles formed in a flowing fluid presents a challenge in space applications. Existing methods of the bubble separation in lab-on-a-chip systems appear to be of limited use for space applications due to different functional requirements. We present a new method for the field and flow driven concentration and removal of bubbles from droplets of aqueous polymer solutions. A fluid droplet is placed on an electrically insulating film and serves as a floating electrode capacitively coupled to external electrodes. Following the application of alternating current voltage, bubbles accumulate on the drop surface. A voltage pulse is then applied to remove part of drop enriched with bubbles.

The work is supported by NASA grant NNX09AK06G.

See more of this Session: Advances In Electrokinetics and Electrophoresis - Fundamentals
See more of this Group/Topical: Topical 3: 2011 Annual Meeting of the American Electrophoresis Society (AES)

(6c) Probing the Basset Force Acting On a Particle Undergoing AC Electrophoresis

Christopher L. Wirth, Paul J. Sides and Dennis C. Prieve, Department of Chemical Engineering, Carnegie Mellon University, Pittsburgh, PA

The electrophoretic motion and related assembly of colloidal particles near an electrode is important to a variety of applications including the preparation of colloidal crystals,[1] the separation of colloids,[2,3] and the high throughput evaluation of electrocatalysts[4,5]. With the aid of Total Internal Reflection Microscopy (TIRM), we studied high frequency (1 kHz - 100 kHz) AC electrophoresis of a single particle normal to an electrode. Surprisingly, the particle adopted a position either closer to- or farther from the nearby wall even when the driving force was oscillatory. The position dependence in the hydrodynamic mobility of the particle suffices to explain apparent repulsion from the electrode, but attraction toward the electrode had remained a mystery for a number of years.[6] Recently, these experiments and new modeling suggested that the Basset force plays a role in this unexpected behavior. The Basset (or "history") force, results from the transient development of the particle's momentum boundary layer during unsteady particle motion. TIRM's capability of probing weak, ~kT scale interaction energies is ideal for investigation of this hydrodynamic effect. The evidence showing the significance of including the Basset force is the subject of this presentation.

1. Joannopoulos, J. D.; Villeneuve, P. R.; Fan, S. H., Photonic crystals: Putting a new twist on light. Nature (1997) *386* (6621).

2. Zhou, H.; White, L. R.; Tilton, R. D., Lateral separation of colloids or cells by dielectrophoresis augmented by AC electroosmosis. Journal of Colloid and Interface Science (2005) *285* (1).

3. Lecuyer, S.; Ristenpart, W. D.; Vincent, O.; Stone, H. A., Electrohydrodynamic size stratification and flow separation of giant vesicles. Applied Physics Letters (2008) *92* (10).

4. Sides, P. J.; Wirth, C. L.; Prieve, D. C., An Imaging Ammeter for Electrochemical Measurements. Electrochemical and Solid State Letters (2010) *13* (8).

5. Wirth, C. L.; Sides, P. J.; Prieve, D. C., The imaging ammeter. Journal of Colloid and Interface Science (2011) *357* (1).

6. Fagan, J. A.; Sides, P. J.; Prieve, D. C., Evidence of multiple electrohydrodynamic forces acting on a colloidal particle near an electrode due to an alternating current electric field. Langmuir (2005) *21* (5).

See more of this Session: Advances In Electrokinetics and Electrophoresis - Fundamentals

(6d) Joule Heating Effects On Fluid and Particle Transport In Insulator-Based Dielectrophoresis Devices

Sriram Sridharan[1], Junjie Zhu[2] and **Xiangchun Xuan**[1], (1)Mechanical Engineering, Clemson University, Clemson, SC, (2)Clemson University, Clemson, SC

Insulator-based dielectrophoresis (iDEP) is an emerging technology that has been widely used to manipulate particles in microfluidic devices. However, due to the locally amplified electric field around the in-channel insulator, Joule heating often becomes an unavoidable issue that may disturb the electroosmotic flow and affect the particle motion. This work presents a combined experimental and numerical study of Joule heating effects on fluid and particle transport in a typical iDEP device, e.g., a constriction microchannel, under DC-biased AC voltages. It is observed that depending on the magnitude of the DC voltage, a pair of counter-rotating fluid circulations can occur at either the downstream end alone or each end of the channel constriction. These circulations are found to affect the dielectrophoretic deflection of particles and thus mitigate the particle focusing and trapping performances in iDEP devices.

See more of this Session: Advances In Electrokinetics and Electrophoresis - Fundamentals
See more of this Group/Topical: Topical 3: 2011 Annual Meeting of the American Electrophoresis Society (AES)

(6e) Impact of Material Morphology On Bioseparations In Nanocomposite Hydrogels: a Modeling Approach In An Annular Channel

Jeffery W. Thompson, Holly A. Stretz and Pedro E. Arce, Chemical Engineering, Tennessee Technological University, Cookeville, TN

Separation of biomacromolecules is important for potential developments in new pharmaceuticals as well as clinical diagnostics. Separation may take place in a hydrogel , for example. New next-generation nanocomposite hydrogels have the potential to alter biomacromolecule separation characteristics. These nanocomposite gels usually consist of a polyamide matrix with some incorporated nanoparticle. The incorporation of anisotropic high aspect ratio nanoparticles, such as montmorillonite, offers the potential to create new geometric domains within the gel, such as an annulus.

A geometric approximation is employed using continuum mechanics over a singular annular channel with included high aspect ratio nanoparticle will be presented. This approach is based upon the electrokinetics-hydrodynamics framework presented by Arce and others. The impact of physicochemical parameters, symmetry, and, perhaps most importantly, geometry on separation of model particles will be presented under situations involving both electrophoresis as well as electroosmotic flow.

See more of this Session: Advances In Electrokinetics and Electrophoresis - Fundamentals
See more of this Group/Topical: Topical 3: 2011 Annual Meeting of the American Electrophoresis Society (AES)

(6f) AC Electrohydrodynamic Flows In Flame Plasmas

Aaron M. Drews[1], Kyle J. M. Bishop[1], Ludovico Cademartiri[2], Robert Shepherd[2], Aaron Mazzeo[2] and George M. Whitesides[3], (1)Chemical Engineering, Penn State University, University Park, PA, (2) Chemistry and Chemical Biology, Harvard University, Cambridge, MA, (3)Harvard University, Cambridge, MA

The combustion of hydrocarbon flames generates significant amounts of charged species including positive and negatively charged ions, electrons, as well as charged carbonaceous particles ("soot"). Importantly, the movement of these species under the influence of an external electric field can induce hydrodynamic flows with the potential to enhance, control, or even extinguish combustion processes. In this context, the effects of static (DC) fields have been well-documented and applied in practice (e.g., to reduce pollutant production by electrostatic precipitation or extend the limits of operation in fuel-lean combustion). By contrast, time-varying (AC) electric fields (typical frequencies <10 kHz) have yet to be exploited and offer several advantages over DC fields such as the frequency-selective manipulation of charged species and the reduction of ion screening effects. Here, we describe the frequency and voltage-dependent response of methane flames subjected to spatially-uniform but time-varying electric fields. The response of the flame is characterized by two methods. AC electrohydrodynamic gas flows resulting from the collisional momentum transfer between charged and neutral species are characterized using schlieren particle image velocimetry. Additionally, the electrical response of the flame plasma is quantified using impedance spectroscopy. The experimental results are explained by a charge transport model which accounts for the production and consumption of charged species within the flame, their movement in the applied field, and their influence on the surrounding fluid. In closing, we highlight emerging opportunities for the application of electrokinetic/electrohydrodynamic concepts to combustion-related applications.

See more of this Session: Advances In Electrokinetics and Electrophoresis - Fundamentals
See more of this Group/Topical: Topical 3: 2011 Annual Meeting of the American Electrophoresis Society (AES)

(6g) Determining Zeta Potential Through the Solution Displacement Method

Sumita Pennathur, Mechanical Engineering, University of California at Santa Barbara, Santa Barbara, CA

The solution displacement method is a common technique used to experimentally determine the zeta potential of substrates for electrokinetic and electrophoretic applications. However, this method requires modification when dealing with inhomogenous substrates, non-uniform geometries, and large concentration differences. In this work we describe a model that can be used in all microfluidic, electrokinetic and electrophloretic applications of the solution displacement method to determine zeta potential.

See more of this Session: Advances In Electrokinetics and Electrophoresis - Fundamentals
See more of this Group/Topical: Topical 3: 2011 Annual Meeting of the American Electrophoresis Society (AES)

Dielectrophoretic assessment of microparticle dielectric properties employing a planar carbon electrode platform

Abstract ID# 223865

Víctor H. Pérez-González,[1,**] *Héctor Moncada-Hernández,*[1,**] *Rodrigo Martínez-Duarte,*[2] *Sergio O. Martínez-Chapa,*[1] *Marc J. Madou,*[3,4] *Blanca H. Lapizco-Encinas,*[5,*]

[1] *BioMEMS Research Chair, Tecnológico de Monterrey, Monterrey, NL, Mexico.*
[2] *Microsystems Laboratory 4, École Polytechnique Fédérale de Lausanne, Lausanne, Switzerland.*
[3] *Mechanical & Aerospace Engineering, University of California, Irvine, Irvine, CA, US.*
[4] *Ulsan National Institute of Science and Technology, World Class University Program, Ulsan 689-79, South Korea*
[5] *Microscale Bioseparations Laboratory and Department of Chemical Engineering, Tennessee Technological University, Cookeville TN, US.*

**Correspondence author. Email: blapizco@tntech.edu*
***Both authors contributed equally to this work*

Abstract

Carbon-Electrode Dielectrophoresis (CarbonDEP) is a novel variation of the conventional electrode-based DEP. The use of carbon, instead of metal, offers advantages like a wider electrochemical stability window, excellent biocompatibility, and great mechanical properties. In order to optimize the performance of electrokinetic experiments, such as separation and concentration of microparticles, it is necessary to know the dielectric properties of the particle of interest a priori. For the extraction of dielectric properties using DEP, it is required that other electrokinetic phenomena, such as electroosmotic flow (EOF) and electrophoresis (EP), are not present on the experiment. Presented in this work is the implementation of a dielectric properties characterization platform based on CarbonDEP. A microdevice containing planar carbon electrodes was employed to manipulate carboxylated polystyrene particles with diameters of 1 μm and 2.28 μm. Particle responses were obtained by varying the magnitude and frequency of the applied AC potential. Velocity induced on the particles was measured experimentally and electric parameters of the microchannel were obtained through simulations in COMSOL Multiphysics. With this information, a system of non linear equations was built, from which the dielectric properties can be extracted. Potential applications of this work include, but are not limited to, environmental screening for water contamination, food safety, clinical analyses, and improvement of clean energy production methods. The results of this work have great potential to be used as guidelines for the further design and operation of CarbonDEP based systems.

Introduction

Microscale electrokinetic methods are becoming increasingly important in many fields; there is a great potential for microdevices to be employed as standard tools in laboratory and industrial practice. Dielectrophoresis (DEP) is an electrokinetic transport mechanism that has been successfully employed to manipulate particles in microfluidic devices. DEP has been implemented to either separate [1, 2], concentrate [3], transport [4], detect [5], or determine dielectric properties of particles [6] using metal-electrode or insulator-based DEP. Traditional DEP devices are based on the metal electrode scheme where planar electrodes are fabricated on the bottom of a microchannel or chamber [7]. Carbon-Electrode Dielectrophoresis (CarbonDEP) is a novel variation of the conventional electrode-based DEP [8]. The use of carbon, instead of metal, offers advantages like a wider electrochemical stability window, excellent biocompatibility, and great mechanical properties. The other leading trend in DEP research is insulator-based DEP (iDEP) where an electric potential, in the order of hundreds of Volts, is applied employing external electrodes in a microchannel or chamber containing insulating structures. The presence of these insulating structures results in a distortion of the electric field, giving rise to large field gradients where DEP occurs [9]. Nonetheless iDEP presents the inconvenient of Joule heating due to the large electric potentials needed to create high field gradient zones. This need also hinders its application outside a research setting and creates a risk of electric shock for the user.

The knowledge of micrometer particles dielectric properties is of great importance in current microtechnology research. When the permittivity and electric conductivity of particles are unknown, it becomes a tedious work to perform separation of particles of interest from within a heterogeneous mixture. A frequency sweep has to be performed in the lab in order to find the dielectrophoretic separation window. Information about dielectric properties allows for a more efficient experimental phase. This not only applies to separation technology, but also for concentration or transport among others.

The work presented herein treats the problem of dielectric properties determination of particles using DEP in a carbon-based microsystem. A microchannel equipped with planar electrodes on its bottom was excited with alternating current (AC) electric potentials, creating a non uniform electric field across the microchannel as a function of the shape and size of the electrodes. Dielectrophoretic spectra of homogeneous polystyrene microparticles were obtained by varying the amplitude and frequency of the AC signal applied to the electrodes. Dielectrophoretic velocity was then estimated based on the videos obtained from the microscope. DEP velocity measurements were finally used to build a set of mathematical equations as functions of the permittivity and electric conductivity of the particles. This set of equations was solved to obtain estimations of dielectric properties of the particles of interest. Many potential applications in fields such as clinical, biomedical, food safety and environmental analysis could be benefited from the availability of dielectric properties data of target cells and bioparticles.

Theory of Dielectrophoresis

Dielectrophoresis is a technique developed by Pohl in 1951 [10]. It describes the polarization effect of a particle suspended in an electrolytic solution when subjected to a nun uniform DC or AC electric field [11-14]. The dielectrophoretic force experienced by a spherical particle is defined as:

$$\mathbf{F}(\omega) = 2\pi\varepsilon_m r^3 \operatorname{Re}[f_{CM}]\nabla E_{rms}^2 \tag{1}$$

where ε_m is the permittivity of the suspending medium, r is the particle radius, and f_{CM} is the Clausius-Mossotti (CM) factor, which is defined as:

$$f_{CM} = \frac{\varepsilon_p^* - \varepsilon_m^*}{\varepsilon_p^* + 2\varepsilon_m^*} = \frac{\sigma_p^* - \sigma_m^*}{\sigma_p^* + 2\sigma_m^*} \tag{2}$$

where $\varepsilon^* = \varepsilon - j\sigma/\omega$ and $\sigma^* = \sigma + j\varepsilon\omega$ refer to the complex permittivity and complex conductivity respectively. In (2), subscripts p and m refer to particle and medium respectively, and ω is the angular frequency of the applied potential. The CM factor is a measure of the ease of polarizability of a particle in a given suspending medium. When the value of the CM factor is positive, it indicates that the DEP force experienced by the particle will be positive or attractive (pDEP). Otherwise, if f_{CM} is negative, the force will be negative or repulsive (nDEP).

The dielectrophoretic velocity, $\vec{u}_{DEP}$, is defined as the product of the squared rms electric field gradient, ∇E_{rms}^2, and the dielectrophoretic mobility:

$$\mu_{DEP} = \frac{\varepsilon_m r^2 \operatorname{Re}\{f_{CM}\} c}{3\eta} \tag{3}$$

where η is the suspending medium viscosity and c is a correction factor that accounts for unconsidered phenomena and numeric errors from the simulations [15]. From the definition of $\vec{u}_{DEP}$ and using equation (3) it is possible to find an expression for the real part of the CM factor.

$$\operatorname{Re}\{f_{CM}\} = \frac{3|u_{DEP}|\eta a}{\varepsilon_m r^2 c \left|\nabla E_{rms}^2\right|}, \quad \begin{cases} a = 1, & \text{pDEP} \\ a = -1, & \text{nDEP} \end{cases} \tag{4}$$

In order to obtain estimators of dielectric properties, an analytical expression for the real part of the CM factor, in terms of the permittivity and electric conductivity, is required. Performing some algebraic manipulation on equation (2) the desired expression is obtained as:

$$\operatorname{Re}\{f_{CM}\} = \frac{\varepsilon_p^2 + \varepsilon_p\varepsilon_m - 2\varepsilon_m^2 + \left(\sigma_p^2 + \sigma_p\sigma_m - 2\sigma_m^2\right)/\omega^2}{\varepsilon_p^2 + 4\varepsilon_p\varepsilon_m + 4\varepsilon_m^2 + \left(\sigma_p^2 + 4\sigma_p\sigma_m + 4\sigma_m^2\right)/\omega^2} \tag{5}$$

Equating equations (4) and (5), and substituting medium parameters, measured dielectrophoretic velocities, and simulated values of ∇E_{rms}^2, for different electric potential frequencies, it is possible to build the required system of equations in function of the dielectric properties of the particles of interest.

Mathematical simulation

Since the dielectrophoretic velocity is directly proportional to the squared rms electric field gradient, it is mandatory to have a clear understanding of the distribution of the electric field inside the microchannel. To achieve this, COMSOL *Multiphysics*© (COMSOL Inc., Burlington, MA) was employed to perform a finite element analysis (FEA) on the geometry.

The geometry of a microfluidic device was captured with COMSOL *Multiphysics*©. The microchannel consisted of interdigitated planar electrodes, which are used as DEP traps. Planar electrodes are 30 μm wide, the gap on the x-direction between electrodes is 45 μm, and the gap on the y-direction is 100 μm. These interdigitated electrodes are connected to an external excitation source (Figure 1).

In addition to this, MATLAB (The Mathworks Inc., Natick, MA) was employed to evaluate the expressions presented in the theoretical section of this manuscript. This requires feeding the algorithm with results obtained from the FEA simulation and with the experimental measurements of dielectrophoretic velocity.

The approach taken in this work is that of measuring dielectrophoretic velocities of the particles while subjected to AC electric potentials of different frequencies. These measurements are then used, jointly with that obtained from COMSOL, to build the set of equations that are finally solved for the permittivity and electric conductivity of the particles. Three different solution-pairs are obtained by solving the set of equations, from which, only one pair exhibit positive values for both dielectric properties.

Experimental setup

A CarbonDEP device with the geometry described above was fabricated by pyrolysis of SU-8 structures defined by a two-step photolithography process following standard C-MEMS techniques. A thin SU-8 protective layer is deposited over the microchannel floor in order to prevent damage to the electrodes. The microfluidic components were fabricated using double-sided pressure-sensitive adhesive and polycarbonate.

DI water with a conductivity of 1 μS/cm was employed as suspending medium. Conductivity was measured with a multi-parameter bench meter (Model HI 255, Hanna Instruments, Woonsocket, RI, USA).

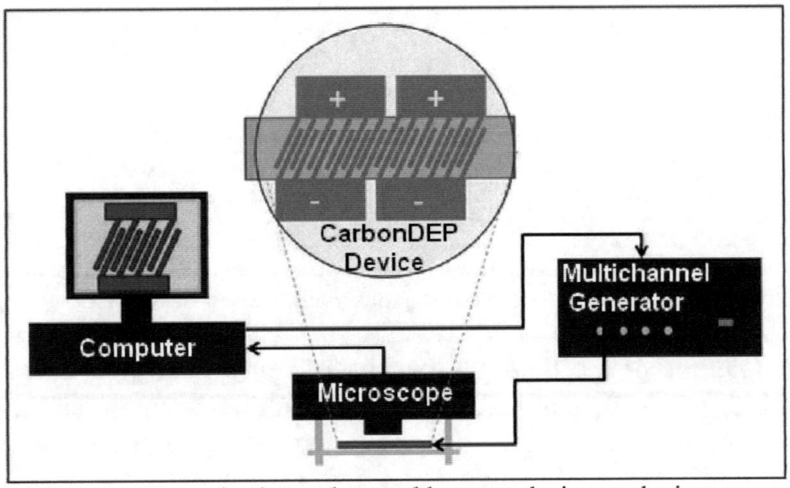

Figure 1. Experimental Set-up, communication and control between the inverted microscope and the multichannel generator with a personal computer.

Carboxylated fluorescent polystyrene particles with a diameter of 2.28 μm color Suncoast Yellow ex/em 540/600 (FC05F, Bangs Laboratories Inc.); and 1 μm color yellow-green fluorescent ex/em 505/515 (F8823, Invitrogen, Or, USA) were employed in this work. Both particle samples were prepared in the suspending medium at the concentration of $1x10^7$ spheres/mL.

The sample containing the microspheres was introduced to the microdevice employing a micropipette and reservoirs were then sealed to avoid pressure gradients affecting particles movement inside the microchannel. The microdevice was mounted under an inverted epifluorescence video microscope for microfluidics SVM340 (LabSmith, Livermore, CA, USA) turned upside down as shown in Figure 1. Electric signal was provided to the microdevice by a high frequency multichannel generator fabricated in-house with frequency range from 1 Hz to 20 MHz and output voltage up to 20 Vpp. A personal computer was employed to manipulate the communication and operation of the microscope and the multichannel generator (Figure 1). Different dielectrophoretic responses were video recorded with the microscope and particles velocities were estimated employing in-house software by evaluating changes in particles position each frame.

Results

As previously mentioned, it is important to study the distribution of the electric field within the microchannel. COMSOL *Multiphysics* was used to perform a simulation to solve for the gradient of the squared rms electric field. In Figure 2, the geometry of the device is shown, as well as the observation window employed to obtain dielectrophoretic velocity measurements. The red line indicates a measurement path from which data was obtained to plot Figure 3.

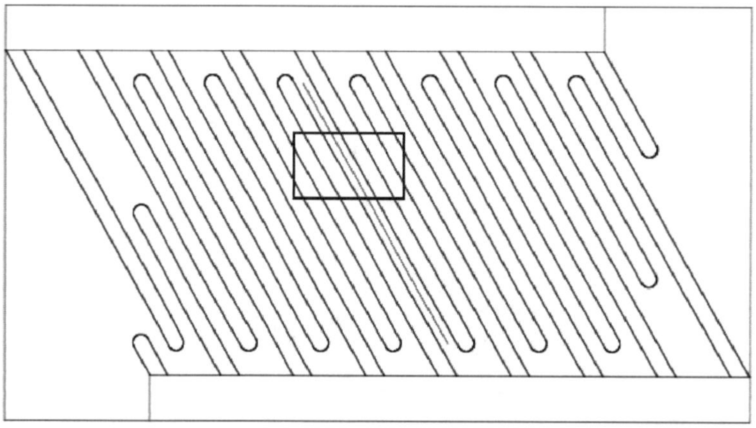

Figure 2. Geometry of the proposed microchannel. The red line indicates one measurement path for the gradient of the squared rms electric field. The blue square shows the observation window for velocity measurements.

Figure 3 shows the magnitude of the field gradient in a logarithmic scale, this makes it easier to read since there are almost four orders of magnitude between the minimum and the maximum. This plot, although is just representative, allows understanding which zones of the channel have the highest magnitude of this field, and which have the lowest. This information was used to approximate the average value of the gradient, which is needed for the estimation of the particles dielectric properties.

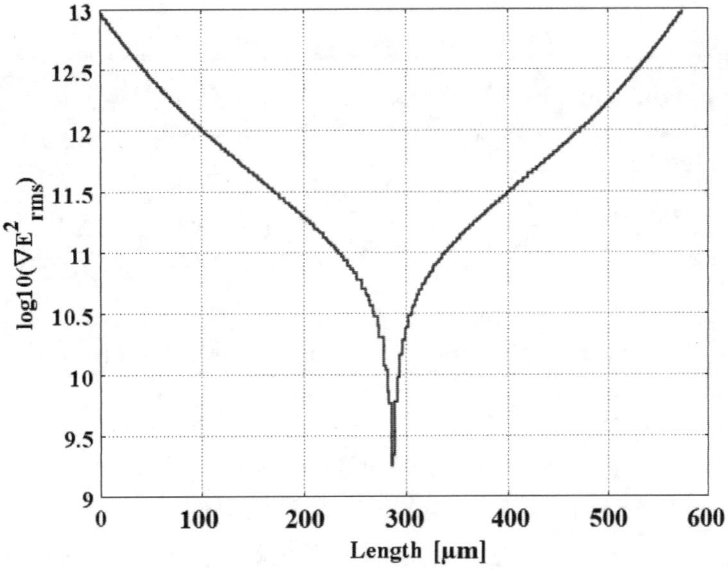

Figure 3. Gradient of the squared rms electric field along the red path shown in Figure 2.

Experimental measurements of dielectrophoretic velocity were obtained through observation of particle displacements over time. A sequence of video frames is shown in Figure 4. Figures 4a-4c are representative of positive DEP based movement, whereas Figures 4d-4f exhibit negative DEP based movement. 2.28 μm polystyrene beads were employed in this Figure. A frequency of 100 kHz was used to induce positive DEP while negative DEP was achieved through the application of a 1 MHz AC signal. The peak to peak value of the AC electric potential applied externally to the electrodes was 15V.

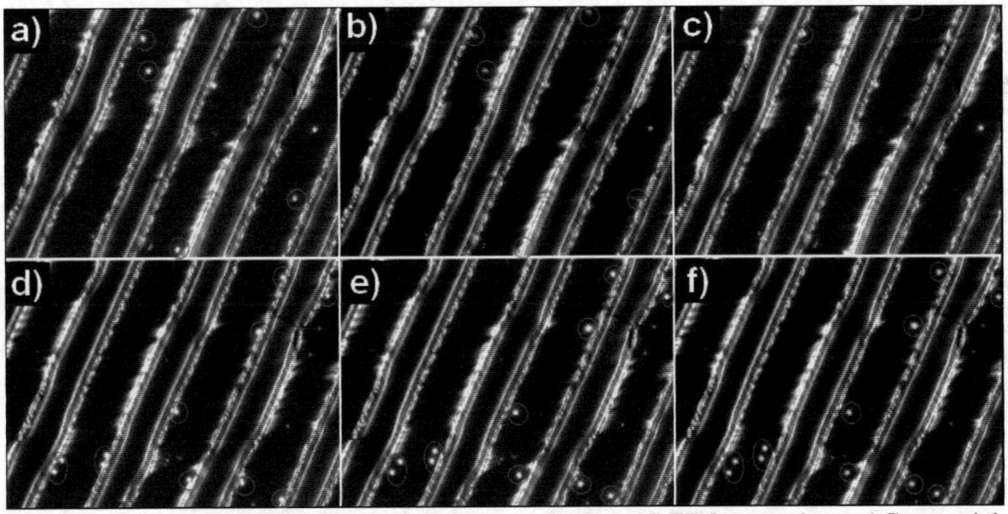

Figure 4. Sequence of 2.28 μm polystyrene particles movement due to DEP force under an AC potential of 15 Vpp. Particles flowing towards the carbon electrodes (pDEP) at 100 kHz: a) initial state, b) after 0.033 s and c) after 0.067 s. Particles flowing away from the carbon electrodes (nDEP) at 1000 kHz: d) initial state, e) after 0.067 s and f) after 0.133 s.

Velocity measurements, as well as a comparison of the theoretical and the experimentally obtained values of the dielectric properties of polystyrene beads are shown in Table 1 below.

Table 1. Summary of properties

Particle Diameter	Electric Potential	Frequency	c	ε_p	σ_p	$\hat{\varepsilon}_p$	$\hat{\sigma}_p$	Theoretical velocity	Measured Velocity
1 µm	20 Vpp	100 kHz	600	2.4	40 µS/cm	16.55	50 µS/cm	107.39 µm/s	111.32 µm/s
		7000 kHz						-58.46 µm/s	-43.62 µm/s
2.28 µm	15 Vpp	100 kHz	387	2.4	18 µS/cm	3.1	19.2 µS/cm	140.88 µm/s	147.29 µm/s
		1000 kHz						-93.73 µm/s	-90.71 µm/s

^ represents estimator.

From these results it can be observed that the estimator shows great similarity in the case of 2.28 µm particles, and a relatively close estimation was also achieved for 1 µm particles. Differences between experimental and theoretical values of velocity and dielectric properties may be due to several reasons; for instance, measurement errors regarding the time frame of the videos recorded, also, particle surface modification, such as carboxyl groups, may affect particle conductivity and permittivity.

Conclusions

A simple methodology for dielectric characterization of microparticles using DEP has been presented. The results presented in this manuscript exhibit the potential of the methodology. There is still important work to do, particularly in the determination of the dielectrophoretic velocity, this in order to improve estimations of permittivity and electric conductivity of the microparticles. Future work includes the extension of this work to multishell particles, such as *E. Coli* and *S. Cerevisiae*.

References

[1] Shafiee, H., Sano, M. B., Henslee, E. A., Caldwell, J. L., Davalos, R. V., *Lab. Chip.* 2010, *10*, 438-445.

[2] Urdaneta, M., Smela, E., *Electrophoresis* 2007, *28*, 3145–3155.

[3] Moncada-Hernández, H., Lapizco-Encinas, B. H., *Anal. Bioanal. Chem.* 2010, *396*, 1805-1816.

[4] Lei, U., Huang, C. W., Chen, J., Yang, C. Y., *et al.*, *Lab. Chip.* 2009, *9*, 1349-1356.

[5] Medoro, G., Manaresi, N., Leonardi, A., Altomare, L., *et al.*, *IEEE Sens. J.* 2003, *3*, 317-325.

[6] Sanchis, A., Brown, A. P., Sancho, M., Martínez, G., *et al.*, *Bioelectromagnetics* 2007, *28*, 393-401.

[7] Jones, T. B., *Electromechanics of Particles*, Cambridge University Press, USA 1995.

[8] Martinez-Duarte, R., Gorkin-Iii, R. A., Abi-Samra, K., Madou, M. J., *Lab. Chip.* 2010, *10*, 1030-1043.

[9] Cummings, E. B., Singh, A. K., *Anal. Chem.* 2003, *75*, 4724-4731.

[10] Pohl, H. A., *J. Appl. Phys.* 1951, *22*, 869-871.

[11] Morgan, H., Green, N. G., *AC electrokinetics : colloids and nanoparticles*, Research Studies Press, Baldock, Hertfordshire, England 2003.

[12] Jones, T. B., *Electromechanics of particles*, Cambridge University Press, Cambridge ; New York 1995.

[13] Green, N. G., Ramos, A., Morgan, H., *J. Phys. D. Appl. Phys.* 2000, *33*, 632-641.

[14] Jones, T. B., *IEEE Eng. Med. Biol.* 2003, *22*, 33-42.

[15] Moncada-Hernandez, H., Baylon-Cardiel, J. L., Pérez-González, V. H., Lapizco-Encinas, B. H., *Electrophoresis* 2011, *32*, In press. DOI: 10.1002/elps.201100168.

(101a) The Effect of Disorder On DNA Dynamics In Post Arrays During Electrophoresis

Daniel W. Olson and Kevin D. Dorfman, Department of Chemical Engineering and Materials Science, University of Minnesota - Twin Cities, Minneapolis, MN

Short DNA (<10 kbp) is commonly separated by size during electrophoresis in disordered, self-assembled gels with pore sizes on the order of 200 nm. Long DNA (~50 kbp) can be separated by a dc electric field when the pore size is several microns. Indeed, separations have been shown in both highly ordered microfabricated post arrays and in disordered quasi-hexagonal matrices of self-assembled magnetic beads. While there have been studies of single molecule dynamics in ordered and disordered systems, the role of disorder is unknown. To study how the disorder in geometry affects DNA separation, we use top-down fabrication techniques to create post arrays with varying degrees of disorder. Using fluorescent microscopy we examine the dynamics of single DNA molecules in these arrays. Additionally, we automate stage motion during experiments to follow single molecules over large distances. We will present data on the probability distributions for the holdup time and distance in arrays of different disorder and compare our results to the existing theories for these devices.

See more of this Session: Microfluidics: Bioanalytical Applications
See more of this Group/Topical: Topical 3: 2011 Annual Meeting of the American Electrophoresis Society (AES)

(101b) Bioparticle Capture In a Sawtooth Dielectrophoretic Microchannel

Paul V. Jones, Sarah J.R. Staton and Mark Hayes, Chemistry and Biochemistry, Arizona State University, Tempe, AZ

Biological fluids can be considered to contain information-rich mixtures of biochemicals and particles that enable clinicians to accurately diagnose a wide range of pathologies. Rapid and inexpensive analysis of blood and other bodily fluids is a topic gaining substantial attention in both science and medicine. One line of development involves microfluidic approaches that provide unique advantages over entrenched technologies, including rapid analysis times, microliter sample and reagent volumes, potentially low cost and practical portability. The present study focuses on the isolation and concentration of bioparticles using DC insulator-based gradient dielectrophoresis in a converging, sawtooth-patterned microchannel. The channel design enables localized isolation and concentration of specific particles by exploiting variations in their characteristic physical properties. Experiments have resulted in successful capture of human blood cells from small-volume samples of diluted whole blood. The technique can operate with isotonic buffers, allowing capture of whole cells. We have also demonstrated dielectrophoretic capture and concentration of mature amyloid protein fibrils. In each case, reproducible capture occurred at specific locales within the sawtooth channel.

See more of this Session: Microfluidics: Bioanalytical Applications
See more of this Group/Topical: Topical 3: 2011 Annual Meeting of the American Electrophoresis Society (AES)

Effects of cell shape on electrokinetic immobilization efficiency in insulator-based dielectrophoretic devices.

Abstract ID#: 224125

*Héctor Moncada-Hernández,[1] Javier Baylon-Cardiel,[1] Victor H. Pérez-Gonzalez,[1] Blanca H. Lapizco-Encinas.[2]**

[1] BioMEMS Research Chair, Tecnológico de Monterrey, Monterrey, NL, Mexico.

[2] Microscale Bioseparations Laboratory, Chemical Engineering Department, Tennessee Technological University, Cookeville TN, US.

**Correspondence author. Email: blapizco@tntech.edu*

Dielectrophoresis (DEP) is defined as the motion of a dielectric particle due to a force as a result from the polarization of the particle in the presence of a non-uniform electric field. DEP has been proven as an efficient electrokinetic technique for the manipulation of particles in microfluidic systems. An alternative to the use of microelectrodes in order to create non-uniform electric fields; is the use of insulating structures along the microdevice between two electrodes in order to "bend" the electric field. This approach is known as insulator-based dielectrophoresis (iDEP). In the present work, the performance of an iDEP microdevice is evaluated employing bioparticles with different geometry shape and size. A computational model is developed to predict and estimate the theoretical trapping zones in the microdevice due to negative DEP response from the particles to evaluate. *Saccharomyces cerevisiae* cells with a spherical shape and *Escherichia coli* cells with a prolate ellipsoidal shape are employed in this work. Experimental and theoretical results are compared in the present study.

Introduction

In the past two decades, advances in microfabrication technologies have allowed the development of miniaturized devices for biological applications. Lab-on-a-chip systems are being employed in a variety of fields; for instance, medical diagnoses, biological assays, food safety, water monitoring, etc. due to their small size and attractive advantages such as: shorter separation times, lower costs, higher performance in terms of selectivity and sensitivity, and smaller required quantities of samples and reagents. These advantages have considerably increased the development of microdevices for the separation of analysis of bioparticles.

With the implementation of dielectrophoresis (DEP), higher selectivity and sensitivity in shorter separation times can be achieved. DEP is defined as the motion of a dielectric particle due to a force as a result from the polarization of the particle in the presence of a non-uniform electric field [1]; and has been proved to be a technique with a great potential for miniaturization. Cummings and Singh proposed the use of arrays of insulating posts between two electrodes so the electric field lines are distorted along the array generating non-uniform electric fields [2]. This technique is known as insulator-based DEP (iDEP). DEP on a particle also depends greatly on its dielectric properties, as well as on its shape and size. Cells are complex particles consisting of several layers of different thickness and dielectric properties [3]. Cell shape can be approximated employing different geometries such as ellipsoids and spheres corresponding to cells as *E. coli* and *S. cerevisiae*, respectively [4]. Dielectric properties of such microorganisms have been reported in different studies [5, 6].

There have been important mathematical modeling efforts to further develop design and performance of insulator-based dielectrophoretic microdevices. In 2003, Cummings and Singh demonstrated experimentally and theoretically the different flow regimes developed by the interaction of diffusion, electrokinetic (EK) and dielectrophoretic flow. These regimes are known as "Streaming DEP" and "Trapping DEP". The first one occurring when DEP overcomes diffusion but EK overcomes DEP; and the later when DEP is able to overcome both diffusion and EK [2]. In 2008, Davalos *et al.* demonstrated that iDEP microdevices may be also produced in a polymer-based platform and achieve a separation and concentration of particles as with glass mircodevices. Also, by employing different surface coatings, performance could be enhanced by lowering the required potential to achieve particle trapping. The theoretical study consisted in showing the possible trapping regions in terms of the ratio between E^2 and the magnitude of the local $\boldsymbol{E}$ within the posts array [7]. Kwon *et al.* described the theoretical background of EK and DEP forces to characterize particle flow in terms of the applied potential in order to propose an improved microdevice to increase particle transport [8]. Baylon-Cardiel *et al.* studied the dielectrophoretic manipulation of particles employing DC and AC signal. They developed a mathematical model to predict the regions where dielectrophoretic immobilization of polystyrene microspheres is achieved in an iDEP microdevice employing DC [9] and AC signal with low frequencies and different waveforms [10].

In the present study, an iDEP microdevice with an array of cylindrical posts was employed to concentrate and separate two types of cells of different shape and size. A computational model was designed and employed to predict the effect on DEP response of the cell size and shape along with the dielectric properties of its membrane, wall and cytoplasm. Concentration and separation of *E. coli* and *S. cerevisiae* cells was experimentally achieved employing DC electric potentials. Mathematical modeling was performed in order to predict the pathlines and magnitude of the regions of dielectrophoretic immobilization for these experiments. In order to understand the mechanism behind DC-iDEP immobilization of biological cells, an analysis of the electrokinetic and dielectrophoretic mobilities was performed.

Theoretical background and mathematical model.

Electrokinetics. The motion of a particle due to the electrical force exerted on it can be expressed in terms of the time dependant EK velocity as:

$$\vec{u}_{EK} = \mu_{EK}\vec{E} \qquad (1)$$

Where μ_{EK} is the electrokinetic mobility and $\vec{E}$ is the spatially non-uniform electric field. The mobility μ_{EK} takes into account the contribution of electrophoretic (EP) motion of the particle and the electroosmotic (EO) motion of the suspending medium.

DEP for spherical particles: the dielectrophoretic force exerted on a spherical particle with a radius R under an electric field with magnitude E is defined as:

$$\vec{F}_{DEP} = 2\pi R^3 \varepsilon_m \mathrm{Re}[f_{CM}]\nabla E^2 \qquad (2)$$

where ε_m is the permittivity of the suspending medium, and f_{CM} is the Classius-Mossotti (CM) factor, which accounts for particle polarizability. The CM factor is defined in terms of complex permittivities, nonetheless, for an applied DC potential, the CM factor can be simplified to [11]:

$$f_{CM} = \frac{\sigma_p - \sigma_m}{\sigma_p + 2\sigma_m} \qquad (3)$$

which is only dependent on the conductivities of the particle and the suspending medium. It is important to observe that eqn. (3) does not have an imaginary part, therefore for the DC case $\mathrm{Re}[f_{CM}] = f_{CM}$.

For a complex particle, such as a cell, the evaluation of the CM factor is not straightforward and a model that accounts for the dielectric and morphological properties of each particle layer must be employed. A simple model for a particle consists of a sphere (the cell cytoplasm, with conductivity σ_{cyto} and radius R_{cyto}), surrounded by two concentric dielectric shells (the cell membrane, with conductivity σ_{mem} and radius R_{mem} and the cell wall, with conductivity σ_{wall} and radius R_{wall}) (Fig 1a). Cell dimensions and dielectric properties employed in this study are shown in detail in Table 1 and Figure 1. In the case of DC electric fields the effective conductivity of the particle that refers to the shell model can be expressed as [12]:

$$\sigma_{eff} = \sigma_{ext}\left[\frac{\left(\frac{R_{ext}}{R_{\mathrm{int}}}\right)^3 + 2\left(\frac{\sigma_{\mathrm{int}} - \sigma_{ext}}{\sigma_{\mathrm{int}} + 2\sigma_{ext}}\right)}{\left(\frac{R_{ext}}{R_{\mathrm{int}}}\right)^3 - \left(\frac{\sigma_{\mathrm{int}} - \sigma_{ext}}{\sigma_{\mathrm{int}} + 2\sigma_{ext}}\right)}\right] \qquad (4)$$

For a 3-shell particle, the previous equation should be used again considering the effective conductivity of the inner shells and the outer shell. Moreover, the contribution of the DEP force to the motion of the particle can be expressed in terms of the dielectrophoretic velocity:

$$\vec{u}_{DEP} = -\mu_{DEP}\nabla E^2 \qquad (5)$$

Table 1. Values of dielectric properties of cells employed for simulations.

Dielectric parameters employed for the model		
	***S. cerevisiae* cells [13]**	***E. coli* Top10F' cells [14]**
Cytoplasm conductivity (σ_{cyto})	0.5 S/m	7x10^{-3} S/m
Membrane conductivity (σ_{mem})	2.5x10^{-7} S/m	1x10^{-6} S/m
Cell wall conductivity (σ_{wall})	1.4x10^{-2} S/m	5x10^{-1} S/m
Cell size	5 μm diameter*	1.5 μm length, 0.75 μm width
Cell membrane thickness (l_{mem})	4.86 nm*	8 nm
Cell wall thickness (l_{wall})	133.6 nm*	50 nm
CM factor	-0.2020	-0.4021
Electrokinetic Mobility (μ_{EK})	2.931x10^{-8} m^2/Vs**	3.335x10^{-8} m^2/Vs**
Dielectrophoretic Mobility (μ_{DEP})	-2.983x10^{-19} m^3/V^{2}s	-3.433x10^{-20} m^3/V^{2}s

* Values were adjusted from [13] according to the size of cells employed in this work.
** Values taken from [15].

where μ_{DEP} is the dielectrophoretic mobility. For a spherical particle of radius R, the mobility $\mu_{DEP,s}$ is defined as:

$$\mu_{DEP,s} = \frac{R^2 \varepsilon_m \operatorname{Re}[f_{CM}]}{6\eta} \qquad (6)$$

where η is the suspending medium viscosity and the subscript "s" refers to spherical particles, where the f_{CM} is given by eqn. (3) and the spherical particle conductivity is given by eqn. (4).

DEP of prolate particles: for a prolate ellipsoidal particle with axes $a_1 > a_2 = a_3$ (see Fig. 1b), where the major axis (a_1) is aligned with the electric field, the dielectrophoretic force exerted can be defined as [16, 17]:

$$\vec{F}_{DEP,pr} = 4\pi a_1 a_2^2 \varepsilon_m \operatorname{Re}(f_{CM}) \nabla E^2 \qquad (7)$$

with a CM factor, for an applied DC electric potential, expressed by:

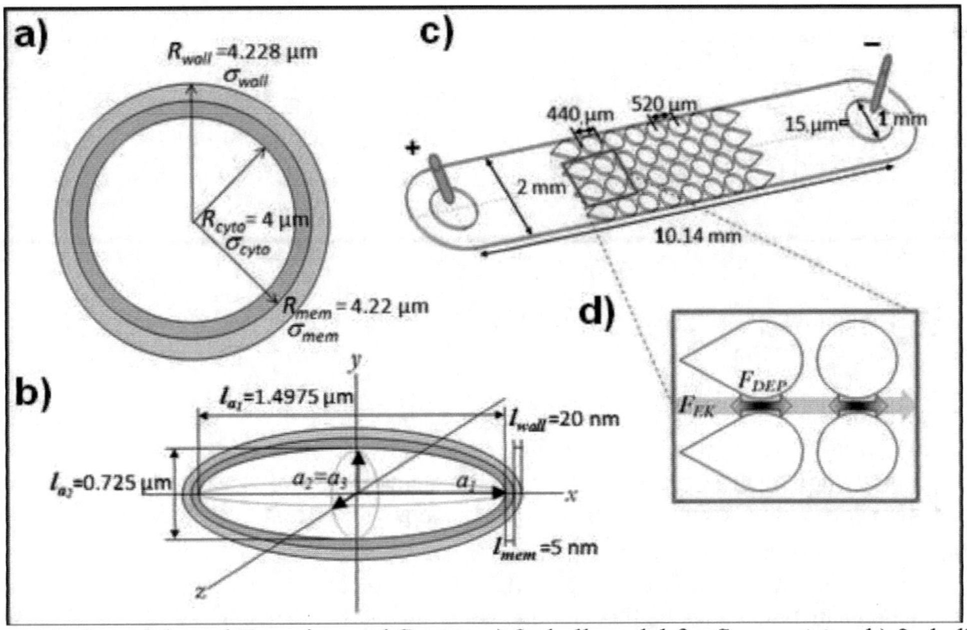

Figure 1. Geometrical models and Experimental Set-up. a) 3-shell model for *S. cerevisiae*, b) 3-shell model for *E. coli*. c) schematic representation of microchannel employed. d) Phenomena affecting the system.

$$f_{CM,i} = \frac{\sigma_p - \sigma_m}{3\left(\sigma_m + \left(\sigma_p - \sigma_m\right)L_i\right)} \quad (8)$$

where the subscript "*pr*" refers to prolate particles, L is the depolarization factor, which leads to three dipole moments, one along each axis of the prolate ellipsoidal particle; the subscript i refers to the x, y and z axis [16, 17]. For this model it is considered that the prolate major axis (a_1) is aligned with the electric field, therefore, there is no net torque exerted on the particle and the dielectrophoretic force can be estimated considering the dipole moment along the main axis of the prolate [17], where the depolarization factor employed is:

$$L_1 = \frac{a_2^2}{2a_1^2 e^3}\left[\ln\left(\frac{1+e}{1-e}\right) - 2e\right] \quad (9)$$

with the following expression for eccentricity:

$$e = \sqrt{1 - \left(\frac{a_2}{a_1}\right)^2} \quad (10)$$

Expressions for ellipsoidal prolate particles considering the multiple layers model were developed for the conductivity of prolate-shaped cells [12, 18]. According to the 2-shell model that comprises the cytoplasm and cell membrane properties, and considering an applied DC electric potential [11], the equivalent conductivity of these layers is defined by the expression:

$$\sigma_{eff} = \sigma_{ext}\left[\frac{\sigma_{\text{int}} + \Delta l\left(\sigma_{\text{int}} + \sigma_{ext}\right)/l_{\text{int}}}{\sigma_{ext} + \Delta l\left(\sigma_{\text{int}} + \sigma_{ext}\right)/l_{\text{int}}}\right] \quad (11)$$

where l refers to the length along the main axis of the prolate particle (a_1), l_{int} is the length of the inner layer along the main axis (a_1), as shown in Figure 1b, and Δl is the length between layers.
As in the case of the sphere model, the expression for the 3-shell prolate model can be obtained by employing this equation again with the outermost layer [12]:
where l_{wall} are the thickness of the cell wall layer (see Table 1). The drag force for a prolate ellipsoidal particle with its major axis aligned with the electric field can be estimated as [17, 19]:

$$\vec{F}_{Drag,pr} = \frac{16\eta\vec{u}\pi a_2^2}{a_2}\frac{1}{\sqrt{1-e^2}}\frac{e}{\left(1+e^2\right)\ln\frac{1+e}{1-e} - 2e} \quad (12)$$

where $\vec{u}$ is the particle velocity. Finally, considering eqns (5), (7) and (12), and assuming a DC electric potential, it is possible to obtain the expression for the dielectrophoretic mobility for a prolate particle with its major axis (a_1) aligned with the electric field [12, 16, 17]:

$$\mu_{DEP,pr} = \frac{a_2^2\varepsilon_m}{12\eta e}\left[\left(1+e^2\right)\ln\frac{1+e}{1-e} - 2e\right]\left[\frac{\sigma_p - \sigma_m}{3\left(\sigma_m + \left(\sigma_p - \sigma_m\right)L_i\right)}\right] \quad (13)$$

where the prolate particle conductivity is given by eqn (11). This equation is equivalent to that derived by Kirby when the condition a_1=$2a_2$ is satisfied [17].

Computational model. The geometry of the iDEP device, schematically presented in Figure 1d, was modeled with *COMSOL Multiphysics* software. A triangular mesh was built in order to solve the Laplace equation using the *Lagrange element* implemented in the software. Finite element analysis was applied in a triangular mesh consisting of 538,500 elements. From the numerical solution it is possible to obtain the distribution of the electric field $\vec{E}$ and the gradient of E^2 inside the array of insulating structures. Dielectrophoretic and electrokinetic velocities, pathlines can be estimated within the microchannel applying concepts of electrokinetics. In this study, it is assumed that the main contributions to the flux of particles $\vec{j}$ along the microchannel come from EK and DEP, as $\vec{j} \approx \vec{u}_{EK} + \vec{u}_{DEP}$ [8, 20]. It can be assumed that particle immobilization is achieved in the regions were the condition $\vec{j} \cdot \vec{E} = 0$ is satisfied. In terms of the EK and DEP mobilities, this condition can be expressed as:

$$(\mu_{EK}\vec{E} - c\mu_{DEP}\nabla E^2)\cdot \vec{E} = 0 \qquad (14)$$

were c is a dimensionless correction factor which accounts for effects of unconsidered phenomena and measurement errors. This expression can be rearranged as a ratio in order to find a condition for particle trapping. Eqn (20) shows this relation [9]:

$$\frac{c\mu_{DEP}\nabla E^2}{\mu_{EK}E^2}\cdot \vec{E} > 1 \qquad (15)$$

By applying these concepts, the mathematical model developed for this study is able to predict the regions where dielectrophoretic immobilization of the cells will occur for a given potential. Electrokinetic mobilities of cells were obtained experimentally in a previous report by our group [15].

Materials and methods.

Escherichia coli Top10F' (PTA-5668, ATCC, VA, USA) were cultured in Luria-Bertani Broth (LB Broth, L3022, Sigma-Aldrich, MO, USA) at 37° C in a shaker incubator for 15 h to achieve late log phase verified by OD measurements at 600 nm. Cells were centrifuged at 7000 rpm for 10 min and the pellet was re-suspended in DI water. Cells were labeled with 3 μL of Syto® 9 bacterial stain (Invitrogen, Carlsbad, CA) that fluoresces green (ex/em 483/503 nm) per mL of cell culture. Labeled cells were centrifuged and washed three times with DI water to remove any excess dye and finally re-suspended in the buffer solution to reach the desired cell concentration ($6\text{x}10^8$ cells/mL).

Saccharomyces cerevisiae (24858, ATCC, VA, USA)were cultured in YM Broth (271120, BD Difco, MD, USA) at 30° C in a shaker incubator for 18 h to achieve late log phase verified by OD measurements at 600 nm. Yeast cells were labeled with Syto® 9 following the same protocol used with *E. coli* cells. After labeling, yeast cells were re-suspended to reach the desired cell concentration ($6\text{x}10^6$ cells/mL).

A 0.5 mM KH_2PO_4 (P2462, DEQ, NL, Mexico) solution in DI water (00068, CTR, NL, Mexico) was prepared as suspending media, resulting in solution with a pH of 7 and a conductivity of 50 μS/cm. The conductivity and pH were measured with a multiparameter bench meter HI 255 (Hanna Instruments, Woonsocket, RI, USA) and a Orion 3 Star Benchtop pH meter (Thermo Scientific, Beverly, MA, USA), respectively.

A schematic representation of the experimental setup is shown in Figure 1c. Experiments were conducted in a microchannel 20 μm deep, 10.12 mm long and 2 mm wide with an array of cylindrical posts (Fig. 1c). The posts have a diameter of 440 μm that are arranged 520 μm center to center and transverse the entire depth of the microchannel.

A high voltage sequencer model HVS448 (LabSmith, Livermore, CA USA) was used to apply DC electric potentials employing platinum-wire electrodes. Visualization of dielectrophoretic behavior of microorganisms was performed employing an inverted epifluorescence video microscope for microfluidics model SVM340 (LabSmith, Livermore, CA USA). The microscope is equipped with a light module with a configuration of blue light LEDs that excites at a wavelength of 480 nm. The filter used in the microscope allows detecting of light above a wavelength of 515 nm. Prior to experimentation, the microchannel was filled with buffer solution and the sample. Afterwards, electrodes were place in the inlet and outlet reservoirs to generate a DC electric field across the microchannel. Dielectrophoretic behavior of the cells was recorded in the form of videos and pictures using the microscope and a personal computer.

Results and discussion.

As discussed in the theoretical background and mathematical model section, three main mechanisms affect the motion of particles in DC-iDEP systems: DEP and the sum of electroosmotic flow and electrophoresis (EK) (Fig 1d). The dielectric properties of cells were introduced to the model to determine the particle conductivity values employing the 3-shell model, which are presented in Table 1.

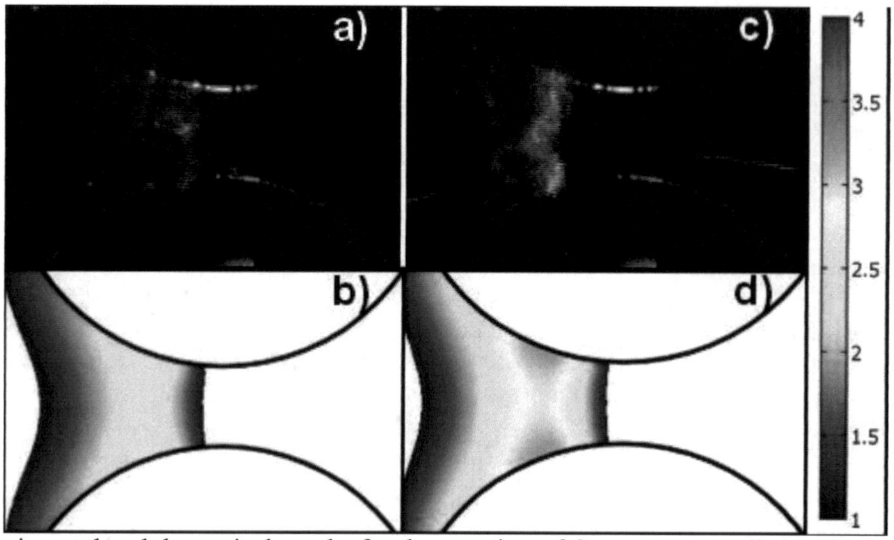

Figure 2. Experimental and theoretical results for the trapping of *S. cerevisiae* cells in a 0.5mM of KH2PO4 solution. a) and b) cells under a potential of 800 V, experimental and simulation results, respectively. c) and d) cells under a potential of 1000 V, experimental and simulation results, respectively.

***S. cerevisiae* cells immobilization.** The behavior of yeast cells was studied through modeling and experimentation. Modeling results are presented in terms of the magnitude of the trapping condition (eqn 15), with a correction factor of 100. Figures 2a and 2b show the experimental results and the estimated trapping zones for yeast cells under a DC potential of ϕ= 800 V. Yeast cells exhibit negative DEP and are being repelled from the zones of higher field gradient, thus, being concentrated in bands just before the constriction created by the insulating posts. When the potential is increased to 1000 V, it can be observed that experimentally well defined bands are formed in the second row of posts (Fig 2c). Also in Figure 2d at ϕ= 1000 V, it is shown that the size of the trapping regions increased and the magnitude of the trapping condition reaches values up to 4 at theconstriction. Trapping zones depicted in Figures 2b and 2d reveal that a trapping condition between 1.5 and 4 is needed to immobilize yeast cells with the present geometry and applied potential. This value is referred to the zone at which cells are completely trapped, the closest to the constriction between the posts (Figs 2a and 2c). It is important to mention that particle interactions are not considered in the mathematical model, and it may be the reason why differences between experimental and predicted results are present.

***E. coli* cells immobilization.** The results obtained for *E. coli* cells are shown in Figure 3, where modeling results are presented in terms of the trapping condition magnitude (eqn 15) employing a correction factor of 500. In Figures 3a and 3b the dielectrophoretic response of the cells with ϕ = 1000 V is observed, these cells were modeled as prolate-shaped particles. As with yeast cells, bacteria cells exhibited negative DEP and were repelled from the zones of higher field gradient. The shape of the trapping region plotted in Figure 3b is in agreement with the experimental results in Figure 3a. When the potential is increased to 1200 V, experimental results in Figure 3c show an even higher cell concentration, and the band of trapped cells increased and shifted to the left. Figure 3d shows that at ϕ = 1200 V, the trapping region changes its shape, increasing in the middle part, which agrees with the experimental results in Figure 3c. Trapping zones showed in Figures 3b and 3d show that the predicted trapping regions agree with the region where bacteria cells are immobilized.

***E. coli* and *S. cerevisiae* cells separation.** By analyzing Figures 2 and 3, it is observed that *S. cerevisiae* and *E. coli* cells exhibit different trapping behavior for a given potential under the present geometry; therefore showing a potential for achieving their separation from a mixture. Experimental and modeling results for the separation of a mixture of *E. coli* and *S. cerevisiae* cells are included in Figure 4. For ϕ = 800 V the experimental results in Figure 4a show that bacterial cells are trapped closer to the constriction than yeast cells, which is expected due to a larger cell size for yeast. By observing the trapped cells in Figure 4a, two different bands of trapped cells are present; to help visualization *E. coli* cells are circled and, since they are smaller than yeast

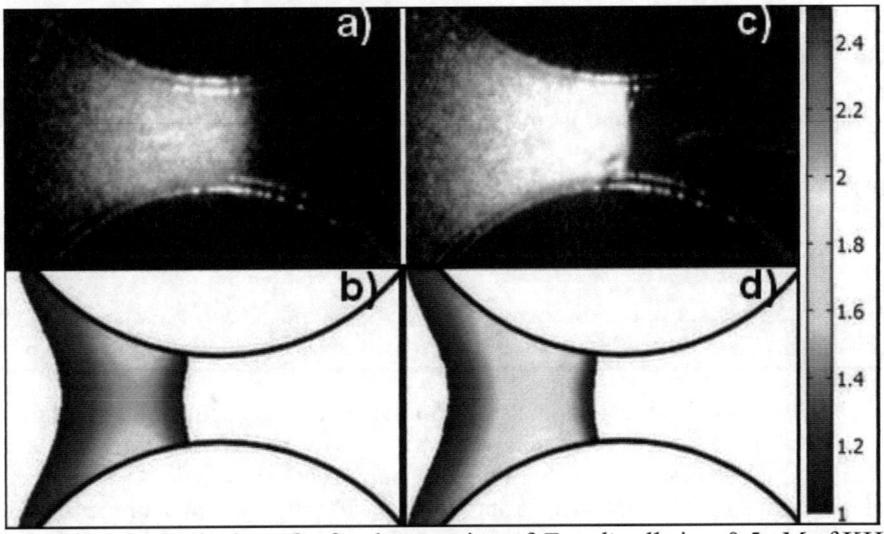

Figure 3. Experimental and theoretical results for the trapping of *E. coli* cells in a 0.5mM of KH2PO4 solution. a) and b) cells under a potential of 1000 V, experimental and simulation results, respectively. c) and d) cells under a potential of 1200 V, experimental and simulation results, respectively.

cells, have a weaker fluorescence and look like a "cloud". Yeast cells are trapped farther away from the constriction, have a stronger fluorescence and look like more defined particles (*i.e.,* not like a cloud). Figure 4b shows the predicted trapping regions for *S. cerevisiae* cells in the upper half and the trapping region for *E. coli* cells in the bottom half at $\phi = 800$ V (same as Fig 4a). As expected, since yeast cells are larger, the magnitude of the trapping region is greater and has a different shape since it extends wider in the y-direction (*i.e.,* perpendicular to the field). As mentioned previously, the differences in shape between *E. coli* and yeast cells (prolate vs. spherical) affect the polarization leading to differences in the form of the trapping regions. Better separation is obtained at $\phi = 1000$ V as shown in Figure 4c, where it is possible to observe two separated bands of cells. Figure 4d shows the trapping regions obtained at $\phi = 1000$ V for yeast (top image) and bacterial (bottom image) cells. It must be noted that particle size and dielectric properties are not the only determinant parameters for achieving

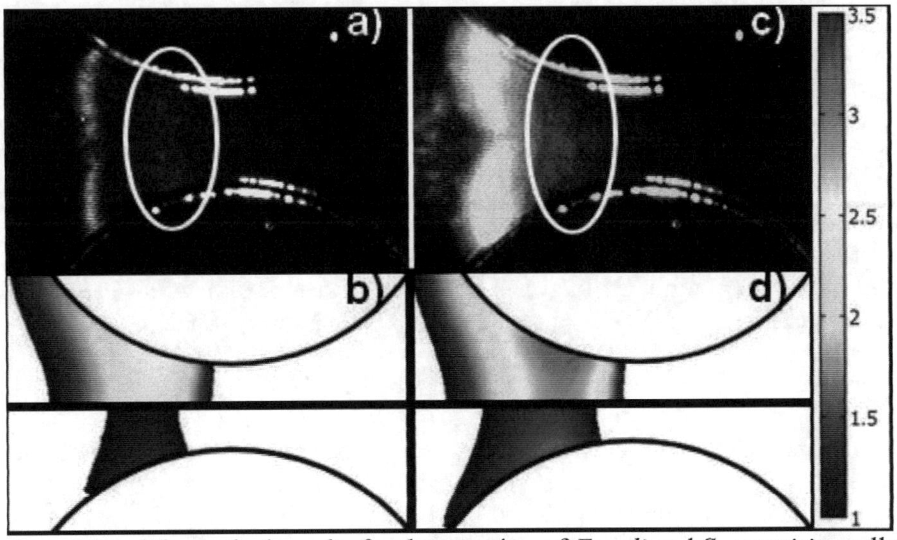

Figure 4. Experimental and theoretical results for the trapping of *E. coli* and *S. cerevisiae* cells in a 0.5mM of KH2PO4 solution. a) experimental separation of cells under a potential of 800 V, b) simulation of the trapping regions for *S. cerevisiae* (top) and *E. coli* (bottom) cells. c) experimental separation of cells under a potential of 1000 V, d) simulation of the trapping regions for *S. cerevisiae* (top) and *E. coli* (bottom) cells.

a neat separation. Particle morphology and even some fluid mechanic effects also affect the system. Therefore, disregarding the fact that yeast cells are much larger than E. coli cells, the distance between separation bands will not be proportional to the difference in size of the particles. These results demonstrate the great potential that mathematical modeling has to offer for the improvement of DC-iDEP systems for cell manipulation. By carefully simulating the dielectrophoretic response and behavior of the cells, microdevices capable of performing complex cell separations and analysis can be designed.

Conclusions

In the present work, the dielectrophoretic behavior of *E. coli* and *S. cerevisiae* cells in a DC insulator-based dielectrophoretic (iDEP) device was studied through mathematical modeling and experimentation. A mathematical model built with COMSOL *Multiphysics* was employed to predict and estimate the squared electric field gradient, dielectrophoretic mobilities, velocity pathlines and the magnitude of the regions of dielectrophoretic trapping for the cells. Experimental work was performed in order to validate the modeling results. It was found that it is possible to predict the dielectrophoretic behavior of cells employing a model that considers the shape of the cell and the dielectric properties of the cell wall, membrane and cytoplasm. The magnitude of electrokinetic and dielectrophoretic mobilities were analyzed and compared, and a condition for dielectrophoretic immobilization was established. The regions for dielectrophoretic immobilization of the cells predicted with the model were in agreement with the experimental results. These results contribute towards the advancement of the dielectrophoretic behavior and characterization of cells under DC electric fields, and will provide with guidelines that can be used in the design of DC-iDEP systems for cell manipulation.

References

[1] Meighan, M. M., Staton, S. J. R., Hayes, M. A., Electrophoresis 2009, 30, 852-865.
[2] Cummings, E. B., Singh, A. K., Anal Chem 2003, 75, 4724-4731.
[3] Jones, T. B., IEEE Eng. Med. Biol. 2003, 22, 33-42.
[4] Morgan, H., Green, N. G., J. Electrostat. 1997, 42, 279-293.
[5] Pethig, R., Markx, G. H., Trends Biotechnol. 1997, 15, 426-432.
[6] Vahey, M. D., Voldman, J., Anal Chem 2009, 81, 2446-2455.
[7] Davalos, R. V., McGraw, G. J., Wallow, T. I., Morales, A. M., Krafcik, K. L., Cummings, E. B., Simmons, B. A., Anal Bioanal Chem 2008, 390, 847-855.
[8] Kwon, J.-S., Maeng, J.-S., Chun, M.-S., Song, S., Microfluid. Nanofluid. 2008, 5, 23-31.
[9] Baylon-Cardiel, J. L., Lapizco-Encinas, B. H., Reyes-Betanzo, C., Chávez-Santoscoy, A. V., Martínez Chapa, S. O., Lab. Chip. 2009, 9, 2896-2901.
[10] Baylon-Cardiel, J. L., Jesús-Pérez, N. M., Chávez-Santoscoy, A. V., Lapizco-Encinas, B. H., Lab. Chip. 2010, 10, 3235-3242.
[11] Markx, G. H., Dyda, P. A., Pethig, R., J. Biotechnol. 1996, 51, 175-180.
[12] Jones, T. B., Electromechanics of Particles, Cambridge University Press, USA 1995.
[13] Urdaneta, M., Smela, E., Electrophoresis 2007, 28, 3145–3155.
[14] Castellarnau, M., Errachid, A., Madrid, C., Juárez, A., Samitier, J., Biophys. J. 2006, 91, 3937-3945.
[15] Garza-García, L. D., Pérez-González, V. H., Pérez-Sánchez, O. A., Lapizco-Encinas, B. H., Chem Eng Technol. 2011, 34, 371-378.
[16] Clarke, R. W., Piper, J. D., Ying, L., Klenerman, D., Phys Rev Lett. 2007, 98, 198102.198101-198102.198104.
[17] Kirby, B., Micro and Nanoscale Fluid Mechanics: Transport in Microfluidic Devices Cambridge University Press New York 2010.
[18] Asami, K., Hanai, T., Koizumi, N., Jpn J Appl Phys. 1980, 19, 359-365.
[19] Fan, F.-G., Ahmadi, G., J. Aerosol Sci. 1995, 26, 813-840.
[20] Cummings, E. B., IEEE Eng. Med. Biol. 2003, 22, 75-84.

(101d) Two-Phase Microfluidic Flow System for the Isolation of Circulating Melanoma Cells Detected Using Photo-Acoustics

Kyle Rood, Christine O'Brien, Rebekah Conley, Thiago Q. DeSouza, John A. Viator and Shramik Sengupta, Biological Engineering, University of Missouri, Columbia, MO

The Photoacoustic Effect occurs when cells that contain pigments (such as melanoma cells) absorb incident laser radiation, suffer a thermo-elastic expansion, and as they expand, give rise to a pressure (acoustic) pulse. By detecting these acoustic pulses using a piezoelectric pressure transducer, we have previously been able to detect very low numbers (< 10/ml) of metastatic melanoma cells, which contain the pigmented granule melanin, in samples also containing much large numbers (> 1,000,000/ml) of leucocytes and platelets, which, because they contain no pigment, are "photo-acoustically transparent". Thus, the Photoacoustic Effect has emerged as a rapid and sensitive method of diagnosing and monitoring Melanoma, the deadliest form of Skin Cancer, at an early stage (before the circulating melanoma cells establish secondary tumors).

Further information about the circulating melanoma cells, such as their molecular biology and gene expression profile can be obtained if these cells can be isolated. But this is a difficult task given their low numbers (< 10/ml) and the high numbers of other cells present (> 1,000,000/ml). We report here, a novel two-phase microfluidic method for isolating our target melanoma cells that is rapid and effective.

When two immiscible fluids (aqueous suspension containing cells, and air, in the present case) are introduced from the two sides of the "horizontal" arm of a T-channel network at constant flow rates and forced to flow together through the same narrow channel (the "vertical" arm of the T-channel), the two fluids, due to the effects of surface tension, spontaneously form alternating "slugs" (droplets that span across the entire cross-section of the channel). Each slug has a volume of ~10 ul, and contains roughly 10^4 WBCs and platelets, but only a very small number of these slugs (approximately equal to the number of target cells) contain target cell(s). The slugs of air present on either side of a slug of aqueous solution confine the acoustic signal, and allows each aqueous slug to be individually interrogated for the presence of photoacoustic- positive (melanoma) cells. At a certain distance downstream, the slugs are sorted, with only the photoacoustically-positive slugs collected, and the rest discarded. This step reduces the number of other cells by a factor of 100. The collected solution can be diluted back to the original volume, and the process repeated multiple times, with a similar 100-fold decrease in the number of other cells, till a suspension containing target cells alone, is obtained. The label-free nature of the isolation technique also eliminates the need for other pre, and post-processing steps that may affect or alter the molecular biology of the targets.

See more of this Session: Microfluidics: Bioanalytical Applications

(101e) Multiplexed Detection of Viral Nucleic Acids In a Combinatorial Microfluidic Screening Chip

Benjamin R. Schudel, **Melikhan Tanyeri**, Arnab Mukherjee, Charles M. Schroeder and Paul J. A. Kenis, Chemical and Biomolecular Engineering, University of Illinois at Urbana-Champaign, Urbana, IL

In this work, we report a multiplexed microfluidic device for detection of nucleic acids for health care applications. Accurate medical diagnosis often requires multiple clinical tests measuring and/or screening for a number of health or disease indicators. At present, however, most diagnostic tests only provide information on a single indicator or marker. For this reason, rapid multiplexed assays based on integrated microdevices could significantly improve the quality of healthcare by reducing turnaround time and analysis cost, and speeding up the decision making process for diagnosis and treatment. Specifically, microfluidic techniques offer several key advantages for diagnostic applications, including miniaturized instrumentation, reduced amounts of sample and reagents for analysis, and integration of assay steps such as sample preparation (processing of physiological fluids), post-processing (e.g. reagent mixing, labeling, separation), detection and analysis into a single device.

Here, we describe a combinatorial screening chip for high sensitivity, sequence-specific screening and detection of viral nucleic acid markers (ssDNA) using surface immobilized molecular beacons and total internal reflection fluorescence microscopy (TIRFM). Using a 4 x 4 array of 200 pL wells, we screened for the presence of four target single stranded oligonucleotides encoding for conserved regions of the genomes of four common viruses: human immunodeficiency virus-1 (HIV-1), human papillomavirus (HPV), Hepatitis A (Hep A) and Hepatitis B (Hep B). Target oligonucleotides are accurately detected and discriminated against alternative oligonucleotides with different sequences. The combinatorial chip represents a versatile platform for the development of clinical diagnostic tests for simultaneous screening, detection and monitoring of a wide range of biological markers of health and disease using minimal sample size.

1. B. R. Schudel, M. Tanyeri, A. Mukherjee, C. M. Schroeder and Paul J. A. Kenis, Lab on a Chip, **11,** (2011).

See more of this Session: Microfluidics: Bioanalytical Applications
See more of this Group/Topical: Topical 3: 2011 Annual Meeting of the American Electrophoresis Society (AES)

(101f) Barcoded Microgel Particles and Scanner for Multiplexed Protein Detection

Rathi L. Srinivas, David C. Appleyard, Stephen C. Chapin and Patrick S. Doyle, Chemical Engineering, Massachusetts Institute of Technology, Cambridge, MA

In the emerging world of medical diagnostics and biological discovery, there is a growing need for highly sensitive and reliable protein detection platforms to make rapid and quantitative measurements in a multiplexed fashion. Using a technology developed in our lab called Stop Flow Lithography (SFL), we are able to generate encoded microgel particles containing one or multiple protein sensing regions. SFL couples the patterning of flows afforded by laminar flow in microfluidic devices with the patterning of light provided by UV lithography. In this manner, we can pattern both the spatial location of probe molecules within a particle (e.g. stripes) and the geometric shape of the particles, giving rise to a barcode. Acrylate modified biological probes are added into the monomer stream and are directly conjugated into the gel matrix. We take advantage of a suspension-based array format to leverage near-solution kinetics and the hydrating environment of a hydrogel scaffold to give proteins a favorable environment for capture and detection. We analyze our particles post-assay using a home-built microfluidic scanning system.

Using both aptamer probes and antibody probes, we have demonstrated capability to detect proteins with high sensitivity and specificity. We initially interfaced our system with DNA aptamer probes against a crucial protein in the blood clotting cascade called thrombin. The aptamer probe we used, HTDQ, is a well characterized thrombin binder that interacts with the heparin binding exosite of the protein. In this set of experiments, the HTDQ-thrombin interaction was studied in the context of hydrogel particles to determine affinity between the probe and target and to ascertain the limit of detection of thrombin. Without the use of additional spacers during immobilization or the need to use multiple aptamer probes to achieve an avidity affect, we were able to show direct detection limits comparable to those in existing microarray platforms.

In tandem, we have done work where we functionalize our particles with monoclonal antibody capture probes. We synthesized particles that could detect il-2, TNF-α, IFN-γ and MIP1-β, which are all cytokines that are highly relevant in the immune response to Human Immunodeficiency Virus -1 (HIV-1). Hybridization assays were initially carried out in 90% fetal bovine serum (FBS), and we were able to detect single pg/mL levels of all cytokines in this complex media without any fouling of the particles. Particles performed equally well in multiplexed settings, showing minimal cross reactivity and high specificity when they were in the presence of several other antibodies; we observed the same detection ranges in multiplexed assays. After initial proof of concept experiments in the FBS, we moved on to analysis of cell culture supernatants derived from peripheral blood mononuclear cells (PBMCs) of HIV-1 infected patients. We were able to discriminate cytokine response from different cell lines, allowing us to profile the response across a range of patients at various disease states.

(101g) Monitoring Pyocyanin Production by Bacteria Using Nanofluidic Electrochemical Sensors

Thaddaeus A. Webster, Anthony P. Fusco, Chun-Wei Kuo and Edgar D. Goluch, Department of Chemical Engineering, Northeastern University, Boston, MA

While processes for manipulating and exploiting bacteria are well known, and have been used for hundreds of years, studies of the interactions between microbial organisms have only begun relatively recently [1]. Of particular interest is the detection of quorum sensing molecules excreted by bacterium that coordinate both inter- and intraspecies interactions. The main objective of this project is to monitor the secretion of the quorum sensing molecule pyocyanin by P. aeruginosa cells as they are exposed to various stressors.

It is known that pyocyanin is an electrochemically active molecule involved in regulating cellular response in its own species, other microbes, and even mammalian cells [2]. We therefore plan to employ nanofluidic electrochemical detection (NED) sensors to detect the local concentration of pyocyanin surrounding microbes while optically monitoring the cellular response. These solid-state devices consist of a 50 nm high solution-filled cavity bounded by two parallel electrodes in which electrochemically active molecules transfer charge between suitably biased electrodes. Electrochemical reactions typically only involve one or a few electrons per molecule rendering direct detection at low concentrations virtually impossible. By utilizing electrochemically active molecules, which can repeatedly undergo reversible reduction and oxidation, each molecule can transfer, on average, thousands of electrons between the electrodes while residing in the cavity [3]. The resulting current is proportional to the concentration of the target molecule.

As a proof of concept, we first tested the ability to detect pyocyanin using commercial redox cycling devices in Phosphate Buffer (PB) and Lysogeny Broth (LB) growth media. Interdigitated electrode array sensors were used to detect pyocyanin concentration in PB and LB in the range of 0.1 – 40 μM. The concentrations detected are physiologically relevant and indicate that sensors based on redox cycling can be used in microbial experiments. Future work will employ NED sensors integrated in microfluidic channels to measure pyocyanin production by P. aeruginosa under various conditions.

References:

[1] Shiner, E.K., et al. (2005) *29*, 935-947

[2] Allen, L., et al. J Immunology (2005) *174*, 3643; Bukelman, O., et al. Chem Commun (2009) 2836.

[3] Goluch, E.D., et al. Anal Bioanal Chem (2009) *394*, 447; Goluch, E.D., et al. The 14th International Conference on Miniaturized Systems for Chemistry and Life Sciences (microTAS) (2010) 482.

See more of this Session: Microfluidics: Bioanalytical Applications
See more of this Group/Topical: Topical 3: 2011 Annual Meeting of the American Electrophoresis Society (AES)

(101h) Scanning Helium Ion Microscope (HIM)-Milled Solid-State Nanopores: Fabrication and Application to Biomolecule Detection

Adam R. Hall[1], Jijin Yang[2], David C. Ferranti[2], Lewis A. Stern[2], Jason Huang[2] and Colin A. Sanford[2], (1)Joint School of Nanoscience and Nanoengineering, University of North Carolina Greensboro, Greensboro, NC, (2)Carl Zeiss NTS, Peabody, MA

The solid-state nanopore is a fast-growing nanofluidic platform for the detection of electrophoretically-driven single molecules. In this presentation, we report the formation of solid-state nanopores using a new method of scanning helium ion microscope (HIM) milling. The fabrication process offers high sample throughput along with fine control over nanopore dimensions, producing single pores with diameters below 4 nm. We will discuss the details and advantages of our fabrication process and demonstrate the utility of the resultant nanopores for biomolecular analysis.

See more of this Session: Microfluidics: Bioanalytical Applications
See more of this Group/Topical: Topical 3: 2011 Annual Meeting of the American Electrophoresis Society (AES)

(179a) Diffusionless Particle Separation In Coherent Arrays of Flow Perturbers

Eric B. Cummings, MaxOut Renewables, Livermore, CA and **Yolanda Fintschenko**, LabSmith, Livermore, CA

The most powerful conventional separation techniques for small molecules are relatively ineffective for particles of practical interest, such as DNA, proteins, viruses, and cells. Diffusion-mediated techniques such as chromatography take impractically long times and open-channel electrophoretic techniques resolve poorly. Random packings, gels, and free-flowing molecular obstacles and flow-perturbers are used effectively to extend the capabilities of electrophoresis, however, our research focuses on answers to the broad question "how can coherence of obstacles or flow pertubers benefit particle separations?" Spatially coherent flow-perturbers can be self-assembled, lithographically patterned, created via optical or acoustic interference, etc. Within a flow channel, they can comprise physical obstacles such as posts and ridges, spatial variations in surface properties such as conductivity or charge density, or spatial variations of fields within the volume, such as optical and acoustic waves. Several new particle separation modalities that employ coherent arraying have already been introduced since the emergence of microfluidics, such as entropic traps, bumpers, and insulator-based dielectrophoresis. These modalities can be generalized as coherently disposed flow perturbers modulating transport from a mobilizing field, such as a pressure or voltage gradient. We explore analytically, numerically, and experimentally a few classes of such "coherent particle scattering" (CPS) based devices and compare them with conventional devices. CPS devices can have substantial theoretical advantages over conventional incoherent or randomly perturbed devices, such as $>10^5$x faster separation speeds and novel resolving abilities. Based on physical structure, potential for rational design, and new separation capabilities, we draw an analogy between CPS vs. conventional devices to diffraction gratings vs. prisms; hologram vs. lenses.

See more of this Session: Plenary Session of the American Electrophoresis Society
See more of this Group/Topical: Topical 3: 2011 Annual Meeting of the American Electrophoresis Society (AES)

(179b) On-Chip Sample Preparation and Nucleic Acid Profiling Using Isotachophoresis

Juan G. Santiago, Mechanical Engineering, Stanford University, Stanford, CA

Robust, automated sample preparation and its seamless integration with other microfluidic functions remains a major unsolved challenge in diagnostic applications. This is particularly true for extraction, purification, and preconcentration of nucleic acids from complex biological samples such as blood, urine, and cultured cells. At Stanford, we are developing on-chip isotachophoresis (ITP) processes which selectively focus a target analyte from a complex mixture, while rejecting unwanted molecules (e. g., proteins or multivalent cations) which may inhibit hybridization or amplification. We will present on-chip extraction of various target nucleic acids including DNA from blood lysate, rRNA from urine lysate, and microRNA from total RNA.

We will also present a novel assay where we simultaneously extract/purify targets, and then identify and quantify specific sequences using molecular beacons. Molecular beacons are short, sequence specific DNA probes which fluoresce upon hybridization. We use ITP to preconcentrate both target and beacons by ~10,000x into order 10 pl reaction volumes. We will show specific and sensitive detection of target sequences in order 2 minutes with minimal sample preparation, and without target amplification. We will demonstrate quantitation of specific sequences of microRNA from total RNA, and of bacterial 16S rRNA from infected urine lysate.

See more of this Session: Plenary Session of the American Electrophoresis Society
See more of this Group/Topical: Topical 3: 2011 Annual Meeting of the American Electrophoresis Society (AES)

(179c) Acoustically-Driven Microcentrifugation

Nick Glass, Richard Shilton, James Friend and **Leslie Y. Yeo**, Micro/Nanophysics Research Laboratory, Monash University, Clayton, VIC, Australia

The Lab-on-a-CD concept has opened up the powerful possibility of carrying out a wide range of microfluidic operations simply by using a compact disk (CD) player to spin a disk on which microfluidic channels and components are fabricated . Nevertheless, the bulk rotation of the entire CD structure is relatively cumbersome, expensive and unreliable, quite the antithesis of microfluidic philosophy. Interconnectivity, i.e., fluid transfer on and off the chip, is also difficult with these devices. Moreover, the rotation of the entire structure is non-specific (i.e., all components on the CD are rotated), which places severe limitations on upstream or downstream components, e.g., the dispensing or detection components, for which rotation may produce undesirable effects. Here, we demonstrate a novel microcentrifugation technique in which symmetry breaking of a planar surface acoustic wave (SAW) propagating along a piezoelectric substrate can generate a rotational acoustic streaming flow in a sessile nanoliter drop (which is several orders of magnitude smaller than the CD) placed atop the substrate. The azimuthal flow is rapid, with linear velocities of several cm/s giving rise to Reynolds numbers between 10 and 100. Such fast azimuthal streaming flows, shown to be chaotic beyond a threshold power, can be used to generate intense micromixing within the drop. Indeed, we show that the rate and yield of a variety of distinct chemical and biochemical reaction classes far exceed that obtained using ultrasonic or microwave-assisted mixing, and with considerably lower power. As an example, in-gel protein sample processing and tryptic digestion can be carried out on-chip in under 30 mins. Further, samples can be atomised from a cheap paper-based microfluidic system, thus demonstrating the potential for direct interfacing with mass spectrometry following chemical synthesis. The microcentrifugation effect can also be exploited for particle manipulation and sorting. We demonstrate this for bioparticle concentration for rapid and sensitive pathogen detection or the separation of red blood cells from plasma for miniaturized diagnostic applications. It is also possible to separate two different particle species by size by exploiting the unique size-dependent scaling between the acoustic and drag forces acting on the particle. Finally, we show the ability of the device to drive the rotation of tiny thin disks made from SU-8 or Mylar down to 100 micron in diameter at speeds close to several thousand rpm.

See more of this Session: Plenary Session of the American Electrophoresis Society
See more of this Group/Topical: Topical 3: 2011 Annual Meeting of the American Electrophoresis Society (AES)

(179d) Subcellular Complexity, An Electrophoretic Perspective

Edgar A. Arriaga[1], Chad Satori[2], **Gregory Wolken**[2], Thane Taylor[2], Jack Doenges[2], Scott Rose[2] and Vratislav Kostal[2], (1)Department of Chemistry, University of Minnesota, Minneapolis, MN, (2) Chemistry, University of Minnesota, Minneapolis, MN

Cell function relies on the coordinated function of thousands of subcellular compartments (organelles). To name a few, mitochondria produce chemical energy in the form of ATP, autophagosomes encapsulate organelles tagged for degradation, and lysosomes degrade and recycle subcellular material. These nanometer-micrometer size organelles have an external membrane made of phospholipids and proteins that are electrically charged at biological pH (7.4), thereby presenting a measurable electrophoretic mobility under biological conditions. Due to the high heterogeneity in size and membrane composition, each organelle type is expected to have a unique electrophoretic profile. As such, each profile would be indicative of organelle subpopulations or would reveal changes that occur in the cell with aging and disease. In order to explore this hypothesis our program has been developing techniques to analyze individual organelles by capillary electrophoresis with laser-induced fluorescence detection (individual organelle CE-LIF).

This presentation will highlight the origins, evolution, current status, and the future of individual organelle CE-LIF in our laboratory. Initial developments based on liposomes demonstrated that under biological conditions electrophoretic mobilities of these organelle mimics are strongly dependent on surface composition. Subsequent developments in individual organelle CE-LIF of mitochondria and lysosomes revealed a high degree of heterogeneity and suggested the existence of electrophoretically heterogeneous subpopulations of these organelles. Individual organelle CE-LIF has also been adequate for measuring individual organelle properties such anti-cancer drug contents, pH, subcellular interactions and production of reactive oxygen species. Ongoing projects include the development of new electrophoretic modes such as isoelectric focusing, immunoanalysis of individual organelles, and theoretical predictions of fundamental properties based on experimental measurements. Long-term, extension of these fundamental principles could lead to novel purification strategies of biological and synthetic particles, adaptation to medical diagnostics, or new tools to explore molecular mechanisms in biomedicine.

See more of this Session: Plenary Session of the American Electrophoresis Society
See more of this Group/Topical: Topical 3: 2011 Annual Meeting of the American Electrophoresis Society (AES)

(179e) Applications of Micro Free Flow Electrophoresis

Michael T. Bowser, Nicholas W. Frost and Meng Jing, Department of Chemistry, University of Minnesota, Minneapolis, MN

In free flow electrophoresis (FFE) a thin stream of sample is continuously introduced into a planar flow chamber. An electric field is applied perpendicularly to the flow through the separation chamber. Analytes are deflected laterally in the electric field according to their electrophoretic mobility giving rise to individual stream paths. FFE has recently been miniaturized into a microfluidic format (uFFE), requiring less sample and reagents, a simplified flow profile and better heat dissipation.

The continuous nature of uFFE separations suggests a number of novel analytical applications. For example, we have demonstrated how introducing a buffer gradient into the uFFE device can be used to efficiently optimize a range of separation conditions or estimate dissociation constants in as little as five minutes. We have also explored introducing a gradient at the sample channel to titrate a fluorescently labeled aptamer with increasing concentrations of its protein target. Due to the continuous nature of gradient uFFE, complete coverage of the binding curve is possible in as little as five minutes. uFFE is an ideal second stage for two dimensional separations. Continuous uFFE separation allows complete sampling of peaks as they migrate off the first dimension column with minimal loss of efficiency. Coupling to uFFE results in a 10-20 fold improvement in peak capacity without sacrificing separation time. uFFE also offers the potential for impressive microscale purification. We will demonstrate isolation of high affinity aptamers using uFFE.

See more of this Session: Plenary Session of the American Electrophoresis Society
See more of this Group/Topical: Topical 3: 2011 Annual Meeting of the American Electrophoresis Society (AES)

(220a) Kinetic Mechanism of Electrochemical Reactions In a Microfluidic Cell

Petr Cervenka, Jiri Hrdlicka, **Michal Pribyl** and Dalimil Snita, Department of Chemical Engineering, Institute of Chemical Technology, Prague, Prague 6, Czech Republic

We propose a novel kinetic mechanism of electrochemical reactions at solid electrode – water electrolyte interface. We assume fast formation of electron donors and electron acceptors (mediators) by thermal fluctuations in the electrode matter. Product of surface concentrations of these mediators is considered to be a constant due to thermal equilibrium. The electron mediators at an electrode can interact with oxidized or reduced forms of electrochemically active substances. In this study, we focus on a simple reversible electrochemical reaction – oxidative dissolving of a metal and its reductive deposition. Two electrodes from the same material immersed into an electrolyte form an electrochemical cell. The electrolyte contains the oxidized form of the metal (monovalent), counter-ion and auxiliary co-ion. The mathematical model consists of molar balances of the ionic components and the Poisson equation. Electric charge concentration on the electrodes is also computed. Concentrations of the electron mediators are then described as functions of the electric charge concentrations. The developed mathematical model enables dynamical and stationary analysis of behavior of the electrochemical cell.

For example, dynamics of the electrochemical cell without electric current load is studied. Dissolution of the electrode metal is observed together with decrease of the electric potential at the electrodes. We also studied behavior of the system under the electric current load. Here we identified a shift of the system from electrochemically limited regime to the transport limited regime. Phenomenon called an "ionic gate" was observed at an electrode. The obtained current-voltage characteristics are in good agreement with predictions given by the classical Frumkin-Butler-Volmer reaction description.

The developed model can be used for investigation various electrochemical systems and microsystems. For example, performance of ac electroosmotic micropumps is significantly affected by electrochemical interactions. Detail and comprehensive studies on this problem is still missing.

See more of this Session: Fundamentals of Electrokinetic Flows: Novel Applications and Ionic Fluxes at Interfaces
See more of this Group/Topical: Topical 3: 2011 Annual Meeting of the American Electrophoresis Society (AES)

(220b) Effect of the Debye Screening Parameter and Electrolyte Valence On the Transport of Ions In Nanochannels

Alexey S. Belyaev and Dimiter N. Petsev, Chemical and Nuclear Engineering, The University of New Mexico, Albuquerque, NM

We examine the effects of electrolyte valence, concentration and pH on the current transport through nanopores in synthetic membranes in the case when the Electric Double Layer (EDL) thicknesses are comparable with dimensions of the nanopores in the membranes. The pores were fabricated by irradiation of the foils with swift heavy ions and subsequent chemical etching. Electrical conductivity measurements showed that the nanopores in the membrane are cation selective and rectify the current with the preferential direction. We present a theoretical and experimental account of the effects of the thicknesses of the electric double layer and pore surface charge on the ionic transport in the nanopores. The analysis leads to better fundamental understanding of the electrical interactions between the fixed and mobile charges of within the nanopore.

See more of this Session: Fundamentals of Electrokinetic Flows: Novel Applications and Ionic Fluxes at Interfaces
See more of this Group/Topical: Topical 3: 2011 Annual Meeting of the American Electrophoresis Society (AES)

(220c) Effect of Morphology On Optimal Separation Times In Nanocomposite Polymer Gel Electrophoresis: Diverging Annular Pore Morphology

Jennifer Anne Pascal[1], Holly A. Stretz[2], Mario Oyanader[2] and Pedro E. Arce[2], (1)Chemical Engineering, Tennessee Tech University, Cookeville, TN, (2)Chemical Engineering, Tennessee Technological University, Cookeville, TN

Experimental efforts have indicated that polymer gels embedded with nanoparticles used in gel electrophoresis, are quite an appealing technology to help improve separation of biomolecules for applications in clinical diagnostics and environmental proteomics, for example [1]. These gels have been used by scientists and engineers to achieve separation of biomacromolecules by taking advantage of their different responses to the applied electrical field, as well as their morphology, or internal structure. The changes in morphology within the gels due to the inclusion of nanoparticles offers a very appealing tuning parameter to improve separation efficiency. These nanocomposite gels can be viewed as a porous media, consisting of capillary channels of different geometries, and if a particular morphology is shown to improve separation efficiency, this morphology could be potentially implemented experimentally.

In this presentation, a pore within the gel of annular morphology with an axially varying cross section and applied orthogonal electrical field will be analyzed using the electrokinetic-hydrodynamics (EKHD) approach to obtain effective transport parameters (i.e. effective velocity and effective diffusivity) that predict the macrotransport behavior of the system. Then, these effective parameters can be used to compute optimal separation times within the annular morphology. Furthermore, the effect of various geometrical parameters and operating parameters on the optimal separation time will be presented.

1. Thompson, J.W., H.A. Stretz, and P.E. Arce, *Preliminary Observations of the Role of Material Morphology on Protein-Electrophoretic Transport in Gold Nanocomposite Hydrogels.* Industrial & Engineering Chemistry Research, 2010. **49**(23): p. 12104-12110.

See more of this Session: Fundamentals of Electrokinetic Flows: Novel Applications and Ionic Fluxes at Interfaces
See more of this Group/Topical: Topical 3: 2011 Annual Meeting of the American Electrophoresis Society (AES)

(220d) Role of Material Morphology On Electrophoresis of Bio-Molecules: Effect of Poiseuille Flow

Jyothirmai J. Simhadri[1], Mario Oyanader[1], Holly A. Stretz[2] and Pedro E. Arce[1], (1)Chemical Engineering, Tennessee Technological University, Cookeville, TN, (2)Department of Chemical Engineering, Tennessee Technological University, Cookeville, TN

Electrophoresis in polymer hydrogels with nanometer-scale pore structure are widely used for the separation and purification of biological macromolecules. In gel-electrophoresis, the internal morphology of the gel also plays an important role in improving the separation. Tuning the nanometer-scale pore structure of the gel either by templating or by adding nanoparticles to improve separations has been the current area of focus. Moreover, analysis of the effects of the nature of the pore cross-section, pore length and diameters on the transport of macromolecules is an important aspect to be studied either analytically or computationally. In this research we propose to computationally analyze different pore models and study the effect of geometry, and Poiseuille flow on the transport of biomolecules in this anisotropic-like media.

See more of this Session: Fundamentals of Electrokinetic Flows: Novel Applications and Ionic Fluxes at Interfaces
See more of this Group/Topical: Topical 3: 2011 Annual Meeting of the American Electrophoresis Society (AES)

(220e) Desalination Shocks In Cross Flows

Ali Mani[1], Daosheng Deng[1], Vicki Dydek[1], Sven Schlumpberger[1] and Martin Bazant[2], (1)Chemical engineering, MIT, Cambridge, MA, (2)Chemical engineering and Mathematics, MIT, Cambridge, MA

Recent experimental and theoretical studies have shown that surface conduction provides an effective mechanism for sustaining overlimiting current in microstructures. It is well understood that passing current through microstructures next to a selective boundary, such as a membrane, gives rise to the concentration polarization phenomenon, through which asymmetric repelling/selection by the boundary initiates a depletion region next to the boundary. For microstructures with high enough surface to volume ratios, such as microporous media or thin microchannels, this initial depletion can induce a long-range nonlinear wave propagating away from the interface faster than diffusion. Propagation of this wave (desalination shock) leaves behind an ultra pure region of bulk electrolyte through which current is dominantly supported by surface conduction. In our presentation we will discuss behavior of desalination shocks in multi-dimensional systems where cross flow, tangential to the membrane, is present acting to distort the shape of the shock. We will discuss advantages of such systems for desalination and water purification.

See more of this Session: Fundamentals of Electrokinetic Flows: Novel Applications and Ionic Fluxes at Interfaces
See more of this Group/Topical: Topical 3: 2011 Annual Meeting of the American Electrophoresis Society (AES)

(220f) Observation of Surface-Charge-Induced Overlimiting Current In Porous Materials

Daosheng Deng[1], Vicki Dydek[1], Ali Mani[1], Sven Schlumpberger[1] and Martin Bazant[2], (1)Chemical engineering, MIT, Cambridge, MA, (2)Chemical engineering and Mathematics, MIT, Cambridge, MA

Salt transport in bulk electrolytes occurs by diffusion and convection, but in microfluidic devices and porous media, the presence of charged side walls leads to additional surface transport mechanisms, surface conduction and electro-osmotic flows, which become more important as the bulk salt concentration decreases. As a result, it is possible to exceed the diffusion-limited current to a membrane or electrode. In this work, we present experimental observations of over-limiting current to an ion-exchange membrane through a porous glass frit with submicron pores. The current-voltage relation and approximately constant over-limiting conductance agree well with theoretical predictions, and high current behavior is consistent with the propagation of "desalination shocks". We also demonstrate the continuous extraction of depleted solution from the frit near the membrane under conditions of over-limiting current. The advantage of working with porous media rather than microfluidic devices is that we can cheaply and easily "scale up" this phenomenon to macroscopic volumes for practical applications to water desalination and purification.

See more of this Session: Fundamentals of Electrokinetic Flows: Novel Applications and Ionic Fluxes at Interfaces
See more of this Group/Topical: Topical 3: 2011 Annual Meeting of the American Electrophoresis Society (AES)

(220g) Electrokinetic Particle Separation by Charge In Spiral Microchannels

Junjie Zhu, Clemson University, Clemson, SC and Xiangchun Xuan, Mechanical Engineering, Clemson University, Clemson, SC

The separation of particles from a heterogeneous mixture is critical in chemical and biological analyses. Many methods have been developed to separate particles in microfluidic devices, the majority of which has been limited to be size based and binary. We present a continuous DC electric field driven separation of carboxyl-coated and non-coated polystyrene beads by charge in a spiral microchannel. This method exploits the inherent electric field gradients formed within the channel turns to manipulate particles by dielectrophoresis. The spiral microchannel is also used to demonstrate a continuous ternary separation of coated and non-coated polystyrene beads by charge and size simultaneously. The observed particle separation processes in different situations are all predicted with reasonable agreements by a numerical model.

See more of this Session: Fundamentals of Electrokinetic Flows: Novel Applications and Ionic Fluxes at Interfaces
See more of this Group/Topical: Topical 3: 2011 Annual Meeting of the American Electrophoresis Society (AES)

(220h) Field Effect Control of DNA Nanoparticle Electrokinetic Translocation Through a Nanopore

Ye Ai[1], **Sang Woo Joo**[2] and Shizhi Qian[1], (1)Department of Mechanical and Aerospace Engineering, Old Dominion University, Norfolk, VA, (2)School of Mechanical Engineering, Yeungnam University, Gyongsan, South Korea

A DNA nanoparticle is electrophoretically driven through a single nanopore connecting two fluid reservoirs on either side under an axial electric field externally imposed. The ionic current is simultaneously altered and measured during the nanoparticle electrokinetic translocation process. Based on the current response to the presence of DNA molecules within a nanopore, one hopes to characterize the nanoparticle such as the determination of each nucleotide base in the DNA nanoparticle. One of the most challenging issues in the nanopore-based DNA sequencing technique is that DNA molecules translocate through the nanopore too fast for detection. Therefore, slowing down the nanoparticle electrokinetic translocation through the nanopore to obtain a detectable current change signal is required. We numerically investigated the field effect control of DNA molecules translocation through a nanopore by a field effect transistor (FET) using a continuum model, composed of the coupled Poisson-Nernst-Planck equations for the ionic mass transport and the Navier-Stokes equations for the hydrodynamic field. The FET contains a gate electrode fabricated on the outer surface of the dielectric nanopore. A potential applied to the gate electrode affects the surface potential on the nanopore's inner wall, which in turn affects the net charge of the ionic electrolyte solution and thus the electroosmotic flow (EOF) inside the nanopore. The particle translocation will be retarded when the induced EOF is opposite to the particle electrophoretic motion. The FET control of the DNA translocation in a nanopore is investigated as a function of the ratio of the particle size to the Debye length, the ratio of the surface charged density of the nanopore to that of the particle, and the dielectric property of the dielectric transistor. Two effects arising from the field effect control, namely the EOF and the particle-nanopore electrostatic interaction, can effectively regulate the DNA translocation through a nanopore.

See more of this Session: Fundamentals of Electrokinetic Flows: Novel Applications and Ionic Fluxes at Interfaces
See more of this Group/Topical: Topical 3: 2011 Annual Meeting of the American Electrophoresis Society (AES)

(237a) A Hybrid Nanoscale Biosensing Platform Based On Dielectrophoresis and Surface Plasmonics

Yunshan Wang, Li-Jing Cheng, Satyajyoti Senapati and H.-C. Chang, Chemical & Biomolecular Engineering, University of Notre Dame, Notre Dame, IN

We report a new platform that combines the advantages of dielectrophoresis (DEP) and Surface Plasmon Resonance (SPR) for pH and molecular sensing. Optical sensing offers numerous advantages over Electrical Impedance Spectroscopy (EIS), as its optical frequency does not alter the conformation of the hybridized target molecules, which presents a major robustness issue with EIS detection. SPR relies on the evanescent optical wave extending just few hundred nanometers above metal surface to sense the presence of target molecules, eliminating the sensing of bulk debris. The platform involves a nanostructure array which can sustain both high electric field and a resonant grating for plasmonics. The high electric field is generated at the nanostructure array with a planar electrode pair and is used to trap target molecules from a flowing solution to the nanostructure. The hydrodynamic shear offered by the large flow (microliter/minute) also allows mismatch discrimination in DNA sensing and minimize non-specific binding in general (Cheng et al, Lab on a Chip, 9, 3193(2009)). Instead of an expensive laser light source at a specific incident angle, efficient coupling by the resonant grating allows the light source to be a broadband Halogen lamp or an embedded LED source. Detection is also done with a miniature spectrometer such that the entire detector footprint is smaller than a brief case. The reflection spectrum of the nano-array exhibits several high-Q distinct peaks at frequencies that are sensitive to the presence of hybridized target molecules within the holes of the array. For pH sensing, we functionalize a single monolayer of poly(2-vinyl-pyridine) on the nanostructure array and exploit how their pH-sensitive persistent length (monolayer height) and refractive index produces a large shift in the resonant frequency. Charging and discharging of counter-ions by the AC field is found to induce a coil-stretch transition and a shift in the Plasmonics resonant peaks. We have developed an ion-condensation strategy to either eliminate this ionic strength sensitivity or strengthen it. Sensitivity and selectivity for a range of sample tests will be reported.

See more of this Session: Microfluidics - Detection
See more of this Group/Topical: Topical 3: 2011 Annual Meeting of the American Electrophoresis Society (AES)

(237b) Mechanistic Study of Biosensing Using Ion Selective Membranes/Gels

Zdenek Slouka[1], Satyajyoti Senapati[2], Christy Ruffing[3], Li-Jing Cheng[1] and Hsueh - Chia Chang[3], (1) Chemical and Biomolecular Engineering, University of Notre Dame, Notre Dame, IN, (2)Chemical & Biomolecular Engineering, University of Notre Dame, Notre Dame, IN, (3)Department of Chemical and Biomolecular Engineering, University of Notre Dame, Notre Dame, IN

We report a mechanistic study of biosensors using an ion selective environment (ISE) as a sensing element. ISE contains a fixed charge (either positive or negative) which enhances/suppresses the flux of counter-ions/co-ions. This imbalance of flux of ions gives rise to formation of a depletion and enrichment region in the electrolyte adjacent to the ISE under DC voltage. Under a sufficiently high electric field, hydrodynamic vortices develop in the depletion region resulting in the mixing with bulk of the electrolyte. This increases the ions concentration in the depletion region (limiting region) that results in the sudden increase of current (over-limiting region). Ion selective membranes, ion selective hydrogels or fluidic nanochannels are examples of these ion selective environments.

When the surface of such an ISE is covered with a layer of biomolecules having an opposite charge, the flux of all ions through this ion selective environment is drastically changed, quite analogous to the nonlinear effects of PN junctions on doped semiconductor IV characteristics. This change can easily be picked up as an electrical signal in measuring current-voltage characteristics, chronoamperometric characteristics or impedance spectrometry. This phenomenon is referred to as charge inversion. We will show that the charge inversion on the ISE results in: (i) suppression of the hydrodynamic vortices due to extended space charge and (ii) water-splitting reaction producing new ions responsible for the observed over-limiting current. The transition from "vortices-controlled" to "water-spitting controlled" over-limiting region is dependent on the degree of coverage of ISE with biomolecules. The detailed description of this new way of sensing technique allows the development of sensitive and specific DNA, RNA and protein biosensors.

See more of this Session: Microfluidics - Detection
See more of this Group/Topical: Topical 3: 2011 Annual Meeting of the American Electrophoresis Society (AES)

Dielectrophoretic separation of particles in a multi-section device

R. C. Gallo-Villanueva[1,2], *V. H. Pérez-González*[1], *R. Davalos*[2] *and B. H. Lapizco-Encinas*[3,*]

[1] *BioMEMS Research Chair, Tecnológico de Monterrey, Monterrey, NL, Mexico.*
[2] *School of Biomedical Engineering and Sciences, Virginia Tech–Wake Forest University, Blacksburg, VA.*
[3] *Microscale Bioseparations Laboratory, Department of Chemical Engineering, Tennessee Technological University, Cookeville, TN.*

**Correspondence author. Email:* blapizco@tntech.edu

Abstract

This work presents a method for simultaneous separation and concentration of a mixture of particles, achieved by means of insulator-based dielectrophoresis. The microdevice used has two sections of insulators to specifically trap targeted particles when a DC electric field is generated. Each part of the device posse a side channel for collection of the separated particles with electroosmotic flow. In this way 1 and 4 μm beads were isolated with 1000 V and then redirected to their respective outlets by applying 300 V in the side channels.

Introduction

Dielectrophoresis (DEP) is the electrokinetic movement of particles due to polarization effects in the presence of non-uniform electric fields[1]. In insulator-based dielectrophoresis (iDEP) regions of low and high electric field intensity, *i.e.*, non-uniformity of electric field, are produced when insulators are present between two electrodes. This technique is increasingly being studied for the manipulation of a wide variety of particles, and novel designs are continuously developed. Despite significant advances in the area, complex mixture separation and sample fractionation continues being one the most important challenges.

In this work, a microchannel design manufactured in PDMS is presented for carrying out DC-iDEP separation of a mixture of particles. The device comprises a main channel with two sections of cylindrical posts with different diameters to create diverse regions of non-uniformity of the electric field, designed for the discrimination and separation of particles of several sizes. Each section of the device has a side channel for the collection of the separated and concentrated particles in a different outlet by electroosmotic flow (EOF). Electric fields were applied in different locations at different times using a high voltage sequencer: i) across the length of the main channel using reservoirs 1 and 2 so particles could be trapped at the two post arrays by negative DEP; and ii) between the side channels using reservoirs 3 and 4 or 5 and 6 to collect the concentrated and separated microspheres by means of EOF.

Theory on dielectrophoresis

The dielectrophoretic force acting on a spherical particle can be defined as following[1]:

$$\vec{F}_{DEP} = 2\pi\varepsilon_m r_p^3 \operatorname{Re}(f_{CM})\nabla E^2 \qquad (1)$$

where ε_m is the permittivity of the suspending medium, r_p is the particle radius, ∇E^2 is the gradient of the squared electric field, and $\operatorname{Re}(f_{CM})$ is the real part of the Clausius Mossotti factor (CM) which accounts for the polarization of the particle in relation with that of the suspending medium and can be approximated to their real conductivities when subjected to low frequency electric fields[2]:

$$\operatorname{Re}(f_{CM}) = \operatorname{Re}\left(\frac{\varepsilon_p^* - \varepsilon_m^*}{\varepsilon_p^* + 2\varepsilon_m^*}\right) \approx \frac{\sigma_p - \sigma_m}{\sigma_p + 2\sigma_m} \qquad (2)$$

where ε *and σ are the complex permittivities and real electric conductivities, respectively, and the subscripts *p* and *m* represent the particle and the suspending medium, respectively. The sign of the CM factor reflects on the response that the particle will show under the non-uniform electric field. If the particle has a higher polarizability than the suspending medium the CM factor will be positive (positive DEP) and it will be attracted to the regions of high gradient of the squared electric field. Otherwise, the CM factor will be negative (negative DEP) and the particle will be rejected from these regions.

The dielectrophoretic mobility (μDEP) and velocity ($\vec{v}_{DEP}$) for a spherical particle are defined as:

$$\mu_{DEP} = \frac{r_p^2 \varepsilon_m f_{CM}}{6\eta} \qquad (3)$$

$$\vec{v}_{DEP} = -\mu_{DEP}\nabla E^2 \qquad (4)$$

When DC electric fields are generated, there are other electrokinetic forces present in the system. The electrokinetic force (EK) is the superposition of electrophoresis (EP) and electroosmotic flow (EOF), and for a microchannel with a negative surface charge, such as PDMS,[3] the electrokinetic particle velocity ($\vec{v}_{EK}$) can be defined as follows[4]:

$$\vec{v}_{EO} = \mu_{EO}\vec{E} \tag{5}$$

$$\vec{v}_{EP} = \mu_{EP}\vec{E} \tag{6}$$

$$\vec{v}_{EK} = \mu_{EK}\vec{E} = (\mu_{EO} + \mu_{EP})\vec{E} \tag{7}$$

where μ_{EO} and μ_{EP} are the electroosmotic and electrophoretic mobilities, and $\vec{v}_{EO}$ and $\vec{v}_{EP}$ are the electroosmotic and electrophoretic velocities, respectively. The electrophoretic and electroosmotic mobilities can be estimated from the Helmholtz–Smoluchowski equation[5]:

$$\mu_{EP} = \frac{\varepsilon_m \zeta_p}{\eta} \tag{8}$$

$$\mu_{EO} = \frac{\varepsilon_m \zeta_s}{\eta} \tag{9}$$

where ζ_p and ζ_s are the zeta potential of the particle and substrate, respectively. As reported by Ermolina and Morgan[5] the zeta potential for 1 μm diameter particles is 0.045 mV, which is around three orders of magnitude lower than the zeta potential reported for PDMS substrate which ranges between 50-100 mV.[6] Therefore, the electrophoretic contribution toward particle motion can be neglected and the following approximations can be made:

$$\mu_{EK} \approx \mu_{EO} \tag{10}$$

$$\vec{v}_{EK} = \mu_{EK}\vec{E} = \mu_{EO}\vec{E} \tag{11}$$

Experimental section

Microdevice description

A microdevice was designed to carry out DC-iDEP experiments and was fabricated in PDMS using soft lithography techniques. The main channel of the device is 2 cm-long, 2 mm-wide, 15 μm deep, and has two sections of insulating structures of different diameters. As it can be seen on Figure 1a, the first segment encloses posts of 437 μm in diameter arranged 487 μm center-to-center, while the posts in the second segment have a diameter of 613 μm and are arranged 653 μm center-to-center. Each section has a side channel for collection of the isolated and concentrated sample by sequentially applying an additional electric field and therefore redirecting the particles to their respective outlets.

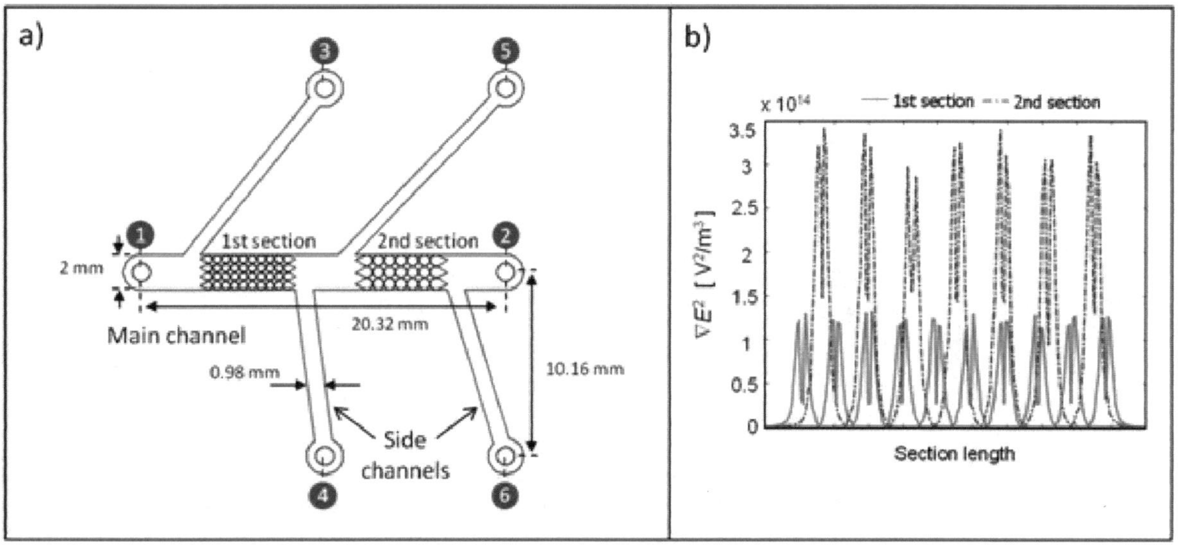

Figure 1. a) Schematic representation of the multi-section channel used for the separation of a mixture of microspheres of 1 and 4 μm, where the first section encloses insulators of 437 μm in diameter arranged 487 μm center-to center while the second section has posts of 613 μm in diameter arranged 613 μm center-to-center; b) gradient of the squared electric field achieved in the first and second section of the device, continuous red and dashed blue lines, respectively.

Dielectrophoretic experiments

Yellow-green carboxylate-modified polystyrene microspheres of 1 μm in diameter, ex/em 505/515 nm, and red sulfate microspheres of 4 μm in diameter (Invitrogen, Carlsbad, CA, USA), ex/em 580/605 nm, were resuspended to a final concentration of $1x10^7$ and $1x10^6$ spheres/mL, respectively. Suspending medium consisted of bidistilled water with a pH of 5.7 and a conductivity of 1.15 μS/cm. After filling the microchannel with the suspending medium, the mixture of particles was feed into the reservoir 1 (see Fig. 1a) and electrical fields were applied between reservoirs 1 and 2 to direct the particles by means of EOF to their particular dielectrophoretic traps. Afterwards, simultaneous electric potentials were applied throughout reservoirs 3 to 4 and 5 to 6 to redirect the separated sample to the collection reservoirs by means of EOF.

Equipment

Dielectrophoretic response of the particles in the multisection channel was observed with the use of a DMI 6000B inverted light microscope equipped with a color camera DFC420, and recorded as pictures or videos processed with the software Leica Application Suite (Leica Microsystems, Bannockburn, IL, USA). 4X and 10X microscope objectives were used for all experiments. DC electric potentials were applied with a high voltage sequencer HVS448, model 300 (Labsmith, Livermore, CA, USA), with the aid of platinum wire electrodes that were placed in the reservoirs (Electron Microscopy Sciences, Hatfield, PA, USA). Both, the inverted microscope and the high voltage sequencer require the use of a personal computer.

Mathematical simulations

All mathematical simulations were performed with the MEMS Module of the software COMSOL *Multiphysics®* (COMSOL, Burlington, MA, USA). The software solves the Laplace equation using Lagrange elements on a triangular mesh that represents the system described in Fig. 1a. The electric potential (ϕ) on the device can be represented by the Laplace equation:

$$\nabla^2\phi = 0 \qquad (12)$$

solved with the following boundary conditions:

$\vec{n}\cdot\vec{J} = 0$ at the boundaries (13)

$\phi = \phi_{in}$ at the inlet of the microchannel (14)

$\phi = 0$ at the outlet of the microchannel (15)

where $\vec{n}$ is the normal vector to the surface, $\vec{J}$ is the electric current and ϕ_{in} is the electrical potential applied between the electrodes. The boundaries include the PDMS surfaces, which consist on the insulating posts and the walls of the channels. In order to achieve immobilization or trapping of particles, the dielectrophoretic component to the particle motion must overcome other transport mechanisms, such as diffusion, pressure driven flow and EK motion. For this work, it was considered that only DEP and EK are the only driving forces with significant effect on the particles. Taking this into account, the model results given by the software include electric potential, electric field, and gradient of the squared electric field.

Results

Simulations performed with the system described in Figure 1a show that the two sections of the microchannel create two different regions of gradient of the squared electric field. Figure 1b represents the magnitude of this value and it can be seen that the second part of the channel, which comprises insulating posts with 613 μm in diameter, generate a value of more than double than that produced by the first section of the device, which has posts of 437 μm in diameter. This value will affect the potential required for DEP to overcome the EK force in each section where, according to Eq. 1, larger particles need a lower dielectrophoretic force to achieve trapping. Therefore, at specific electric potential larger particles will experience dielectrophoretic trapping in the first section of the insulators while smaller particles present will continue their EK trajectories through the channel. When these smaller particles reach the second section of posts, DEP will be strong enough to overcome EK forces since this region produces a higher gradient of the squared electric field, *i.e.*, dielectrophoretic trapping of the particles will be achieved.

Figure 2 shows the experimental results of a mixture of microspheres of 1 (green) and 4 μm (red) in diameter when subjected to an electric potential of 1000 V between reservoirs 1 and 2. Particles were dielectrophoretically separated and concentrated at the same time in their respective section according to size. Figure 2a is a picture of the first region of the channel and it can be observed that only the 4 μm microspheres are being trapped and therefore concentrated while the 1 μm particles are dominated by EOF. In the second section, DEP is high enough to overcome EK forces for the 1 μm spheres but no red particles are detected since they were collected in the previous section of the device (Figure 2b). It can be observed that there is a clear rejection from the regions of high gradient of the squared electric field which is in the narrow part between the posts. This rejection allows trapping when DEP is strong enough to overcome EK forces, and can be strengthen by increasing the electric potential applied. With this electric potential simultaneous concentration and separation of the sample was achieved.

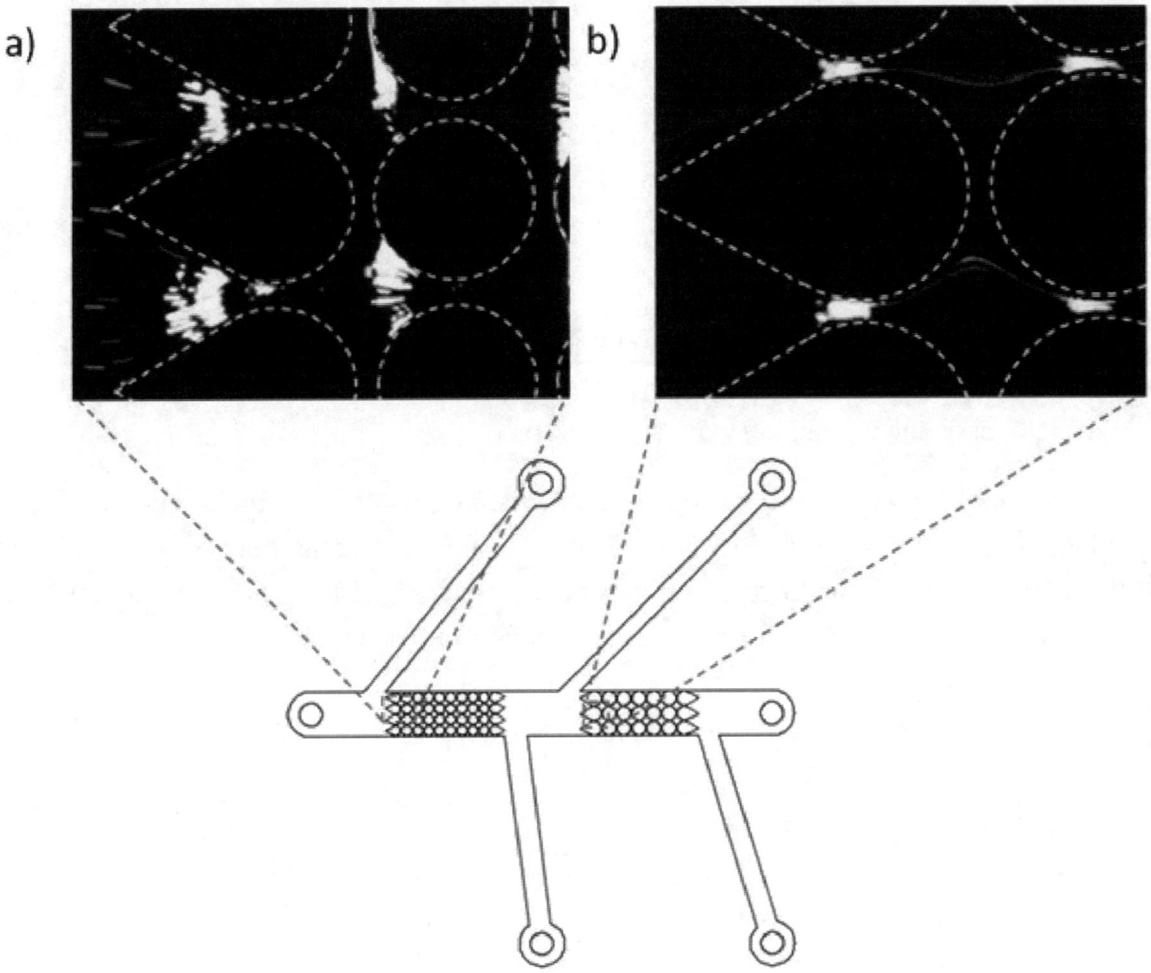

Figure 2. Simultaneous separation and concentration of 1 μm (green) and 4 μm (red) microspheres with an electric potential of 1000 V. a) The 4 μm beads are collected by negative DEP in the first section of the device while the 1 μm beads are dominated by EOF; b) in the second section of the main channel the DEP force is large enough to overcome EOF and therefore trap and concentrate the 1 μm spheres. These results were obtained with a suspending buffer of 1.15 μS/cm and a pH of 5.7.

The system allowed recovery of the isolated and concentration sample by applying a synchronized electric potential through reservoirs 3 to 4 and 5 to 6, while turning of the electric field generated between reservoirs 1 and 2. A simultaneous potential of 300 V was enough to redirect the sample to the collection reservoirs without showing significant DEP to prevent flow.

Concluding remarks

A system was presented that could concentrate particles in a specific region of a microdevice and therefore isolate a mixture of two different microspheres. In addition by using DC electric fields the sample can be directed either to the dielectrophoretic trapping regions if the electric potential is applied between the main inlet and outlet, or to the collection outlets if it is applied in the side channel reservoirs.

Negative DEP was observed for the microspheres, *i.e.*, particles were concentrated before the narrow parts between the posts (regions of high gradient of the squared electric field) showing a rejection from this zones.

These results demonstrate the flexibility of DC-iDEP as a technique where by combining different designs it is possible to achieve separation and individual collection of particles from a mixture. This work provides guidelines for the development and design of microdevices to be employed for the sorting, concentration and collection of particles by means of DC-iDEP.

References

1. Pohl, H. A., *J. Appl. Phys.* 1951, *22*, 869-871.
2. Markx, G. H., Dyda, P. A., Pethig, R., *J. Biotechnol.* 1996, *51*, 175-180.
3. Sze, A., Erickson, D., Ren, L., Li, D., *J. Colloid Interface Sci.* 2003, *261*, 402-410.
4. Martínez-López, J. I., Moncada-Hernández, H., Baylon-Cardiel, J. L., Martínez-Chapa, S. O., Rito-Palomares, M., Lapizco-Encinas, B. H., *Anal. Bioanal. Chem.* 2009, *394*, 293-302.
5. Ermolina, I., Morgan, H., *J. Colloid Interface Sci.* 2005, *285*, 419–428.
6. Buszewski, B., Szumski, M., Klodzinska, E., Dahm, H., *J. Sep. Sci.* 2003, *26*, 1045-1049.

(237d) Dielectrophoretic Cell Sorters for High Purity and Throughput

Guiren Wang, Mechanical Engineering, University of South Carolina, Columbia, SC, Fang Yang, Department of Mechanical Engineering & Biomedical Engineering Program, University of South Carolina, Columbia, SC, Xiaoming Yang, WJB Dorn Veterans Affairs Medical Center, Columbia, South Carolina, Columbia, SC and Hong Jiang, GraceFlow Technology, Irmo, SC, Columbia, SC

Microfluidics by nature deals normally with low throughput. In practical applications such as clinical diagnostics or environmental detection, sometimes relatively large sample handling is required. In sample preparation, high purity of sample is often demanded. Here we present two new microfluidic cell sorters based on negative dielectrophoresis (DEP) for continuous flow operations. One is a cascade configuration sorter designed to increase purity of isolated target cell. And the other one has two staggered side channels in opposite side walls to increase sample throughput without compromising enrichment factor. We first used Particles (carboxylate microspheres) of different sizes to demonstrate the feasibility of the present DEP sorters for cell isolation. Then biological cells, i.e. human prostate cancer cell line LNCaP and human colorectal cancer cell line HCT116 were used to test the performance of the DEP sorters. Comparing to a single side channel DEP cell sorter, the isolation purity was improved from 80% to 96% by single cascade sorter and the sample throughput was increased from 0.2 μL/min to 0.65 μL/min by a single staggered side channel sorter.

See more of this Session: Microfluidics - Detection
See more of this Group/Topical: Topical 3: 2011 Annual Meeting of the American Electrophoresis Society (AES)

(237e) Dielectrophoretic Filter for Engine Oil

Yueyang Shen, Ezinwa Elele and **Boris Khusid**, Chemical, Biological & Pharmaceutical Engineering, New Jersey Institute of Technology, Newark, NJ

We report on a novel concept of an AC dielectrophoretic filter with a three-dimensional electrode array. A filter is constructed by winding into layers around the core tube two sheets of woven metal wire-mesh with several sheets of woven insulating wire-mesh sandwiched in between. Contrary to conventional dielectrophoretic devices, the proposed design of electrodes generates a high-gradient field over a large working volume by applying several hundred volts at a standard frequency of 60 Hz. The operating principle of filtration is based on our recently developed method of AC dielectrophoretic gating for microfluidics. The proof-of-concept is tested by measuring the single-pass performance of two filters on positively polarized particles dispersed in engine oil: spherical glass beads, fused aluminum oxide powder, and silicon metal powder, all smaller than the mesh opening.

See more of this Session: Microfluidics - Detection
See more of this Group/Topical: Topical 3: 2011 Annual Meeting of the American Electrophoresis Society (AES)

(237f) pH Gradient Formation In An Insulator-Based Dielectrophoresis Device Used In Protein Trapping Applications

Aytug Gencoglu[1], Fernanda Camacho-Alanis[2], Vi Thanh Nguyen[2], Asuka Nakano[2], Alexandra Ros[2] and Adrienne Minerick[1], (1)Department of Chemical Engineering, Michigan Technological University, Houghton, MI, (2)Department of Chemistry and Biochemistry, Arizona State University, Tempe, AZ

pH Gradient Formation in an Insulator-Based Dielectrophoresis Device Used in Protein Trapping Applications

Aytu• Geno•lu[1], Fernanda Camacho-Alanis[2], Vi Thanh Nguyen[2], Asuka Nakano[2], Alexandra Ros[2], Adrienne Minerick[1]

1: Department of Chemical Engineering, Michigan Technological University, 1400 Townsend Drive, Houghton, MI, 49931 USA

2: Department of Chemistry and Biochemistry, Arizona State University, Tempe, AZ, 85287-1604 USA

Insulator-based direct current (DC) dielectrophoretic (iDEP) microdevices have the potential to replace traditional alternating current (AC) dielectrophoretic devices for many cellular and biomolecular separation applications. These microdevices employ large DC fields, under which the electrode reactions and electrokinetic ion transport mechanisms become significant enough to affect ion distributions in the nanoliters of fluid in the microdevice. The most commonly encountered of these electrode reactions is the electrolysis of water. With most electrode materials, when an electric field is applied, H^+ and OH^- ions are generated at anode and cathode surfaces, respectively. Both of these ions have high diffusional and electrophoretic mobilities. Therefore, diffusional and electrokinetic transport of these species can be significant enough in micro- and nanoscale systems that H^+ and OH^- concentration gradients can develop within the microfluidic channel. This leads to what is termed a "natural pH gradient," and can be exploited for applications such as isoelectric focusing (IEF). However, natural pH gradients can also cause unexpected fluid behavior in micro- or nanofluidic systems by causing spatial changes, such as nonuniform wall surface charges. Moreover, the properties of analytes may be dependent on pH and the formation of natural pH gradients may not be desirable.

This work shows the formation of natural pH gradients in an iDEP microdevice with pt wire electrodes, under conditions applied during iDEP protein manipulation experiments. pH changes were observed by measuring the fluorescence intensities of pH sensitive dye FITC Isomer I and the pH insensitive dye TRITC and correlating the FITC/TRITC fluorescence intensity ratio to pH. A dependence of natural pH gradient formation on the phosphate buffer solution concentration was observed under 100 V/cm electric fields. When the channel was filled with relatively low concentration solutions (Conductivity: s=0.01S/m or 0.05 S/m), pH was found to drop below 4 in 5 to 10 minutes. However, this behavior was more consistent in the case of 0.01 S/m solution. pH gradient formation was not observed in the case of a relatively high concentration solution (s=0.10 S/m), due to the higher buffering capacity of the solution. pH was observed to drop dramatically in seconds under 3000 V/cm electric fields with all buffer solutions. It was also observed that when the Pt wire electrodes had been in use for more than 1 hour, pH gradient formation did not occur under 100 V/cm electric fields, even with the 0.01 S/m buffer solution. However, 3000 V/cm electric fields still caused rapid pH changes in the microchannels when the used Pt wire electrodes were employed, regardless of the solution concentration.

This work shows that pH gradients can form in iDEP devices, possibly affecting the operation of such devices. It is shown that pH gradient formation is influenced by electric field strength, buffering capacity of the medium, and the properties of the electrode surface. Based on this work, pH gradient formation can be prevented by using solutions with high buffering capacity, by passivating the electrodes prior to use, or by using electrode materials on which water electrolysis reactions do not occur.

See more of this Session: Microfluidics - Detection
See more of this Group/Topical: Topical 3: 2011 Annual Meeting of the American Electrophoresis Society (AES)

(237g) Real-Time Detection of Nanoparticle Deposition by Measurement of Zeta Potential

Paul Sides, Reza Rock and Sairam Shekhar, Chemical Engineering, Carnegie Mellon University, Pittsburgh, PA

Deposition of nanoparticles on planar substrates was followed in real time by monitoring the streaming potential of mica in the presence of alumina nanoparticles near neutral pH. Freshly cleaved mica was affixed to the end of a spindle and rotated in dilute aqueous potassium nitrate. A steady streaming potential was measured. Then, alumina nanoparticles were injected into the solution; the streaming potential rose from negative values characteristic of mica to positive values characteristic of alumina at pH 5.6. The streaming potentials reached steady values on a time scale of a few minutes. Conversion of the streaming potentials to zeta potentials, with the aid of published theory, [1] gave typical values first for mica and then for alumina under the conditions of the experiment. AFM photographs of a rinsed and dried sample showed the presence of particles with a surface coverage around ½. The reciprocal experiment, with silica particles adsorbing on sapphire, showed the expected charge reversal from positive values characteristic of sapphire at low pH to negative values characteristic of silica.

In other experiments, addition of alumina to solution at pH 10 did not change the zeta potential of mica. Lower the pH below the isoelectric point of alumina, however, did induce depositon. Thus, one can determine the isoelectric point of arbitrary particles by charging them negatively at high pH and then lowering the pH until they adsorb at the isolectric point. Indeed, titrating positive particles adsorbed on a negative substrate below their isoelectric point and titrating the same particles on a positive substrate above their isoelectric point might be a method for determining the full titration curve of particles.

The work of Adamczyk and co-workers [2] has established a relationship between surface coverage of particles and zeta potential; the application of this theory to the measured zeta potentials revealed the dependence of the surface coverage of particles as a function of time. According to this theory, surface coverages of ½ or more made the contribution of the substrate to the measured streaming potential essentially vanish. Thus, the theory agrees with the experimental findings that titration of a substrate coated with an adsorbed layer of particles should yield values of zeta potential characteristic of the particles at coverages substantially lower than unity.

Other experiments demonstrated that initial rates of adsorption followed the classical dependence on the square root of rotation rate expected for the rotating disk.

1. P. Sides, J. Newman, J. Hoggard, D. Prieve, *Langmuir* **22** 9765-9769 (2006).

2. Z. Adamcyzk, M. Nattich, M. Zaucha, *Current Opinion in Colloid and Interface Science* **15** 175 – 183 (2010).

See more of this Session: Microfluidics - Detection
See more of this Group/Topical: Topical 3: 2011 Annual Meeting of the American Electrophoresis Society (AES)

(276a) Time-Resolved Measurement Used In Flow Cytometry

Ruofan Cao, Patrick Jenkins and Jessica Houston, Chemical Engineering, New Mexico State University, Las Cruces, NM

Time-resolved measurement used in flow cytometry

R. Cao, P. Jenkins, and J. P. Houston

Department of Chemical Engineering

New Mexico State University

Las Cruces, NM 88003

Time-resolved measurements are widely used in fluorescence spectroscopy, particularly for studies of biological macromolecules and increasingly for cellular imaging. Time-resolved measurements contain more information than is available from the steady-state data. For example, with time-dependent information the kinetic phenomena of biological macromolecules can be obtained as well as interactions of macromolecules with each other (i.e. measurement of Forster resonance energy transfer). In this contribution we present frequency-domain fluorescence measurements from a home-built time-resolved cytometry system (TRCS). This device is capable of measuring decay-kinetic phenomena from fluorescent species in flowing cells or particles, which pass a laser excitation source at microsecond transit times. A quantum dot (QDot) experiment was performed that validated the resolution limits and multiple-lifetime capabilities of the TRCS. QDots were selected because of their favorable fluorescence decay, extremely long Stoke's shift and high photostability. The QDots were bound to the membrane of Chinese Hamster Ovary (CHO-K1) cell populations as well as the surface of polystyrene microspheres. Three primary measurements were made: (1) the capture of average fluorescence lifetimes at different laser modulation frequencies, (2) phase filtering of the QDot fluorescence from other background fluorescence when the QDot was quenched with a heavy metal or non-quenched, and (3) multiple fluorescence decay of the QDot with other organic fluorophores with a novel multi-frequency flow cytometry capability. Many challenges remain before our TRCS can be translated for commercial exploitation. By implementing straightforward analysis of QDots we not only assess and optimize our device such that it becomes simple, stable and efficient, but also delve into more quantitative biological studies that leverage the fluorescence lifetime for true concentration, fraction of molecules in specific environments, monitoring of single cells for kinetic processes, and the elimination of background autofluorescence.

See more of this Session: Biomedical Diagnostics
See more of this Group/Topical: Topical 3: 2011 Annual Meeting of the American Electrophoresis Society (AES)

(276b) Rapid Determination of the Minimum Inhibitory Concentrations (MIC values) of Antibiotics Using a Rapid, Culture-Based, Electrical Detection Method

Sachidevi Puttaswamy, Byung-Doo Lee and Shramik Sengupta, Biological Engineering, University of Missouri, Columbia, MO

For reasons both scientific (such as to study the development of antibiotic resistance in bacteria isolated from different sources) and clinical (such as to ascertain the efficacy of the medication and dose prescribed for a bacterial infection, and make changes if required), one often needs to evaluate the antibiotic susceptibility profile of various bacteria. The susceptibility of a given strain/isolate to a particular antibiotic is usually expressed as the Minimum Inhibitory Concentration (MIC) of the antibiotic.

In practice, a strain of interest is inoculated into liquid cultures with various concentrations of the candidate antibiotic (ranging from 0.5 to 128 ug/ml), and one seeks to determine the minimum antibiotic concentration for which bacterial growth is inhibited. This concentration is known as the Minimum Inhibitory Concentration. Current detection systems typically rely on optical density measurements to determine the occurrence of bacterial growth (or lack thereof) for the presence of given concentrations of antibiotic. While reliable, they take a long time to detect changes in bacterial numbers (and hence determine MIC values). Another major drawback of the existing system is that they cannot distinguish between bacteriostatic and bactericidal action of the antibiotic against specific bacteria.

This work presents a novel electrical method for rapid antibiotic susceptibility testing of bacterial strains of interest. The principle underlying our method of detection is the polarizability of viable bacterial cells. In the presence of an alternating electric field, there occurs a build-up of charge at the membrane, causing the cells to act like capacitors. If however, the cell membrane is compromised (as happens when cells die), this phenomenon no-longer occurs.

In our method, we set up a panel of cultures with different concentrations of the candidate antibiotic (as with current methods). However, each culture is monitored electrically. If the bacteria multiply in number, there will be a corresponding increase in the charge stored in the interior of the suspension (its "bulk capacitance"), and this increase in bulk capacitance indicates that the bacteria are proliferating despite the presence of the antibiotic. Similarly, as the bacterial number decreases due to the bactericidal effect of antibiotics, there will be a corresponding decrease in the value of the bulk capacitance. And finally, if the bacterial numbers are held static by the antibiotic, the bulk capacitance will also remain virtually unchanged.

Thus our detection method has the ability to differentiate between bactericidal and bacteriostatic action of antibiotic for a given bacteria. Using our method we were able to measure the Minimum Inhibitory Concentrations (MICs) of various antibiotics for bacterial strains of interest in <4 hours (as opposed to more than 1 day using current methods).

See more of this Session: Biomedical Diagnostics
See more of this Group/Topical: Topical 3: 2011 Annual Meeting of the American Electrophoresis Society (AES)

(276c) Dependence of Membrane Antigen Expression On Cross-Over Frequency of Human Erythrocytes and Corresponding Theoretical Explanation

Kaela M. Leonard and Adrienne Minerick, Department of Chemical Engineering, Michigan Technological University, Houghton, MI

Since Pohl's initial exploration of dielectrophoretic forceto explain the behavior of polarizable particles in a non-uniform electricfield, adjustments have been made to the Clausius-Mossotti factor to includenon-spherical particles and particles of a core-shell composition. This work has become increasinglyimportant since dielectrophoresis has been applied to biological cell systemsbecause it allows for the separation and sorting of cells of similar origin butslightly varying dielectric properties. Knowing the frequency at which the Clausius-Mossotti factor is equal tozero allows the researcher to determine the conductivity and permittivity ofthe cell of interest. However,current equations describing the Clausius-Mossotti factor ignore slightmembrane differences such as antigen expression despite previous resultsshowing that small membrane changes make appreciable changes to theClausius-Mossotti factor. Thiswork builds upon the previous adjustments made to the dielectrophoretic forceequation to account for not only the shelled ellipsoidal shape of humanerythrocytes but also the differences in membrane antigen expression. The common ABO-Rh typing systemconsists of 4 different membrane antigens and provides the basis for thesestudies. The Rhesus factor (Rh) isa group of transmembrane antigens that indicate positive blood type (presenceof D antigen) or negative blood type (absence of D antigen). The ABO blood grouping system consistsof three surface antigens, whose presence or absence dictate blood type. A and B antigen chains differ from theO backbone by a single sugar molecule, however this small difference getsmagnified because there are 1.5 million copies of the antigen on theerythrocyte surface. In order todetermine the affect of the ABO antigens on the dielectrophoretic behavior,erythrocytes were treated with β(1-3) galactosidase, which cleaves polysaccharide antigens at a galactose bond. Human erythrocytes of known ABO-Rhblood type were suspended in a 0.1S/m dextrose buffer solution and subjected toan electric field of 0.1V/μ m and varying frequency from 100kHzto 80MHz in order to determine their cross-over frequencies. Experiments were repeated with β(1-3) galactosidase treated erythrocytes. It wasfound that the lower cross-over frequency is ABO bloodtype dependent and that galactosidase modification causes a shift in thecross-over frequency. The highercross-over frequency was Rh blood type dependent and did not change after β(1-3) galactosidase modification. The change in the lower cross-over frequency post-modification proved that thepolysaccharide antigen chain influences the dielectric properties of theerythrocyte and therefore the dielectrophoretic behavior. The shelled ellipsoidalClausius-Mossotti factor model was fit to each ABO blood type response toquantify membrane conductivity differences due to antigen expression. This work will provide insight intomolecular level contributions to conductivity and permittivity –drivencell polarizations in dielectrophoretic fields.

See more of this Session: Biomedical Diagnostics

See more of this Group/Topical: Topical 3: 2011 Annual Meeting of the American Electrophoresis Society (AES)

Dielectrophoretic Signature of Mouse Ovarian Surface Epithelium Cells

Alireza Salmanzadeh,[1,2] Harsha Kittur,[1] Roberto C. Gallo-Villanueva,[1,3] Mike B. Sano,[1] Blanca H. Lapizco-Encinas,[4] Mark A. Stremler,[2] P. Christopher Roberts,[5] Eva M. Schmelz,[5] and Rafael V. Davalos[1,2,]*

Email:davalos@vt.edu

[1]School of Biomedical Engineering and Sciences, Virginia Tech-Wake Forest University, Blacksburg, VA

[2]Department of Engineering Science and Mechanics, Virginia Tech, Blacksburg, VA

[3]BioMEMS Research Lab, Tecnologico de Monterrey, Monterrey, Mexico

[4]Microscale Bioseparations Laboratory, Tennessee Technological University, Cookeville, TN

[5]Department of Human Nutrition, Foods, and Exercise, Virginia Tech, Blacksburg, VA

Abstract

In this study we have investigated the dielectrophoretic properties of early and late stages of Mouse Ovarian Surface Epithelium (MOSE) cells, and macrophages. Utilizing contactless dielectrophoresis (cDEP), a relatively new method to dielectrophoretically manipulate cells, we showed that at an applied frequency, these cells trap at different voltages. Moreover, the effect of different geometrical parameters, such as the pillars' arrangement, the size of each pillar, and gap in between pillars are investigated computationally.

Introduction

Ovarian cancer is the leading cause of gynecologic cancer deaths in women in the United States. Every year, over 20,000 women are diagnosed with this cancer, and over 15,000 women die of the disease [1]. The 5-year survival rate for women diagnosed in the early stages is 70-90%, but that for women diagnosed in the advanced stages is less than 30%. 80-90% of the patients with advanced disease do initially respond to chemotherapy, but 85-90% will relapse, resulting in a survival rate of 20-25%. Yet, symptoms of ovarian cancer typically do not arise until it is in the late stages, so detection methods are currently being developed to enable diagnosis in its early stages.

A challenge in current screening techniques is that invasive methods such as laparoscopy and laparotomy are necessary. Due to the low incidence of this cancer (40 per 1,000,000 per year) [2], the danger of surgical complications for false-positive screenings outweighs the advantages for early detection in true-positive women, so screening test specificity must be a minimum of 99.6% [3]. In addition to this, another challenge is to improve sensitivity so that ovarian cancer can be detected even before symptoms arise.

The most popular biomarker for study in screening methods is CA-125. Elevated levels of this biomarker are found in 50% of early stage patients and 80% of late stage patients [4]. Additional techniques include the use of ultrasound to detect morphological changes in the ovary, as well as color-flow Doppler imaging to track changes in blood-flow patterns. Attempts to combine these techniques do improve sensitivity, but somewhat decrease specificity.

Miniaturized systems represent an attractive alternative for clinical analysis due to portability, rapid response time and reduced requirement of sample and reagents. Dielectrophoresis (DEP), a non-destructive electrokinetic transport mechanism, is one of the most employed techniques for the manipulation of cells in microdevices. Traditionally, DEP utilizes arrays of electrodes to produce non-uniform electric fields to polarize particles and induce their electrokinetic movement. Although this approach to apply DEP has been a very successful technique for the manipulation, separation and enrichment of bioparticles [5, 6], it has some difficulties such as bubble formation (due to electrolysis), electrode fouling and delamination, and sample contamination [7, 8]. Modified DEP techniques, such as insulator-based dielectrophoresis (iDEP) have been developed to overcome drawbacks of conventional DEP.

In iDEP, insulating structures are straddled between two electrodes to cause regions of high and low electric field intensity (non-uniformity of the electric field) and therefore induce polarization of the particles [7]. Such systems were proposed by Cummings and Singh [7] and have been applied successfully for manipulation of DNA [8], separation of live and dead cells [9, 10], and separation of cell types.

In this study the possible application of contactless dielectrophoresis (cDEP) for the separation and concentration of cells is presented. In cDEP, an electric field is generated in the fluidic channel by using two side channels which are filled with a conductive fluid. These side channels are separated from the main channels by thin insulating barriers. Not using metal electrodes eliminates some of the conventional DEP disadvantages, such as sample contamination and bubble formation, and resolves issues associated with joule heating [11, 12]. Using high frequency cDEP microdevices, the bandwidth of trapping of these cells was investigated. The results demonstrated that cDEP techniques can offer a great potential for cancer diagnosis.

Theory

The time-average of DEP force acting on a spherical particle in a non-uniform electric field is given by [15]

$$\vec{F}_{DEP} = 2\pi\varepsilon_m r^3 Re\{\frac{\varepsilon_p^* - \varepsilon_m^*}{\varepsilon_p^* + 2\varepsilon_m^*}\}\nabla(\vec{E}_{rms} \cdot \vec{E}_{rms}), \qquad (1)$$

where ε and ε^* are the real and complex permittivity of the suspending of the particle p or suspending medium m, r is the radius of the particle, and $\vec{E}_{rms}$ is the root mean square of the electric field. The complex permittivity is defined as $\varepsilon^* = \varepsilon - \frac{j\sigma}{\omega}$, where ε and σ are the real permittivity and conductivity, $j^2 = -1$, and ω is the radial frequency of the electric field. $\frac{\varepsilon_p^* - \varepsilon_m^*}{\varepsilon_p^* + 2\varepsilon_m^*}$ has been defined as the Clausius-Mossotti factor, K [15]. The magnitude of K can vary from -0.5 to 1. Depending on the sign of K, the DEP force can either be positive, in which case it is directed towards the high electric field gradient region, or negative, in which case it is directed towards the low electric field gradient region. This dependence is related to the polarizability of the particle with respect to the surrounding medium. The frequency at which the DEP force is zero is called the crossover frequency, f_{xo}, *i.e.*, the frequency at which the real part of K changes sign, $Re\{K(\varepsilon_p^*, \varepsilon_m^*)\} = 0$.

As particles move under the influence of the DEP force, they interact with the surrounding fluid and experience a hydrodynamic force. This hydrodynamic force can be approximated as the drag, which is given by

$$\vec{F}_{Drag} = 6\eta r\pi(u_p - u_f), \quad (2)$$

where r is the particle radius and u_p and u_f are the velocities of the particle and medium, respectively.

Materials and Methods

Device Layout

Figure 1 shows the top view schematic of the microdevice. Electrode channels, which are approximately 1 cm long, are separated from the fluidic channel by 20 μm barriers. Several insulating pillars, 100 μm in diameter, have been used in the fluidic channel to increase the nonuniformity of the electric field and enhance DEP force. The size and distance between these pillars has been optimized using computational modeling.

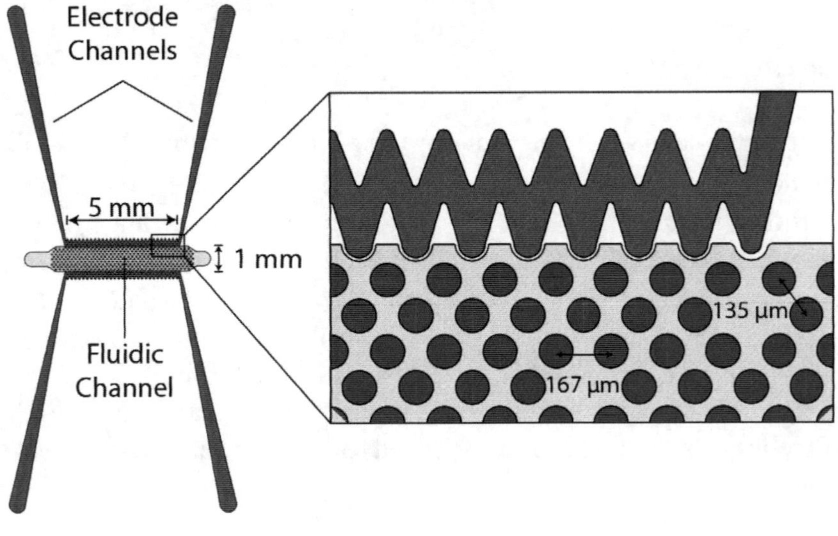

Figure 1. 2D top view schematic of the microdevices. Each pillar is 100 μm in diameter.

Cell Preparation

The samples of early and late stage MOSE cells [16, 17], and PC1 macrophages independently were washed twice and resuspended in a buffer used for DEP experiments (8.5% sucrose [wt/vol], 0.3% glucose [wt/vol], and 0.725% [vol/vol] RPMI) [18] to 106 cells mL-1. The electrical conductivity of the sample was 105±3 μS/cm which was determined using a conductivity meter (Horiba B-173 Twin Conductivity/Salinity Pocket Testers, Cole-Parmer) before each experiment.

Results and Discussion

Computational modeling

Computational models were used to investigate the effect of different geometrical parameters, such as pillar arrangements, the gap between the pillars, and pillar diameter, on the gradient of the squared electric field distribution. We have also studied the fluid flow to predict velocity and shear rate to ensure that the cells' lyses limit is not exceeded.

Our microdevices are made from Polydimethylsiloxane (PDMS) that is bonded to a glass slide. The electrical properties of PDMS have been reported as $\sigma_{PDMS} = 0.83 \times 10^{-12}$ S/m and $\varepsilon_{PDMS} = 2.65$. The electrical conductivities of PBS and DEP buffer are $\sigma_{PBS} = 1.4$ S/m and $\sigma_{DEP} = 0.01$ S/m, respectively, and their relative permittivities are $\varepsilon_{PBS} = \varepsilon_{DEP} = 80$. The electrical properties of PBS and DEP buffer are used for the side and main microfluidic channels, respectively.

The insulating structures inside the main channel are essential tools to enhance the dielectrophoretic force by increasing the electrical resistance of the main channel as well as increasing the nonuniformity of the squared electric field gradient. Without these structures, the gradient of the electric field is limited to areas close to the side walls of the main channel and most of the cells far from the side walls do not experience any dielectrophoretic force. Adding the insulating structures can provide wider channels with uniform gradient of the squared electric field, which consequently, will increase the throughput and selectivity of the cDEP devices. In this study we have used circular pillars.

To compare microdevices with different circular pillars' geometries, a diagonal line located on the center of device has been drawn (line A-B on Fig. 2). Line plots of $\nabla(\vec{E}_{rms} \cdot \vec{E}_{rms})$ inside the main microfluidic channel for two different pillars arrangements, triangular (staggered) and rectangular arrangements, are shown in Fig. 2(b), which demonstrates that triangular arrangement creates higher $\nabla(\vec{E}_{rms} \cdot \vec{E}_{rms})$. Moreover, it has been shown that the calculated hydrodynamic efficiency for capturing cells is better in triangular arrangement than in rectangular[19]. Finally, triangular arrangement increases the likelihood of cell trapping, since there is a higher frequency of cell collisions with the pillars. Thus, a triangular arrangement has been used in our microdevices and we only focused on optimizing the other geometrical parameters for this arrangement.

Circular pillars with different diameters were modeled computationally, and $\nabla(\vec{E}_{rms} \cdot \vec{E}_{rms})$ was calculated on the A-B diagonal line as presented in Figure 2(c). This figure shows that 100 μm pillars allow the generation of higher DEP force than 50 and 200 μm pillars. Figure 2(d) is representing the line plot of the gradient of the electric field for three devices with 100 μm diameter and different gaps between the circular posts (30 μm, 60 μm, and 120 μm). These results show that the gradient of the squared electric field has an inverse relation with the distance between the posts, *i.e.*, the DEP force decreases by increasing the gap between the posts, although this increase is not significant. However, this distance cannot be less than 30 μm as such devices will mechanically filter the cells due to their size. Thus, microdevice with pillars with 30 μm gap has been used in this study. Decreasing the gap between pillars also increases the number of pillars per unit area, which increases the chance of cell trapping.

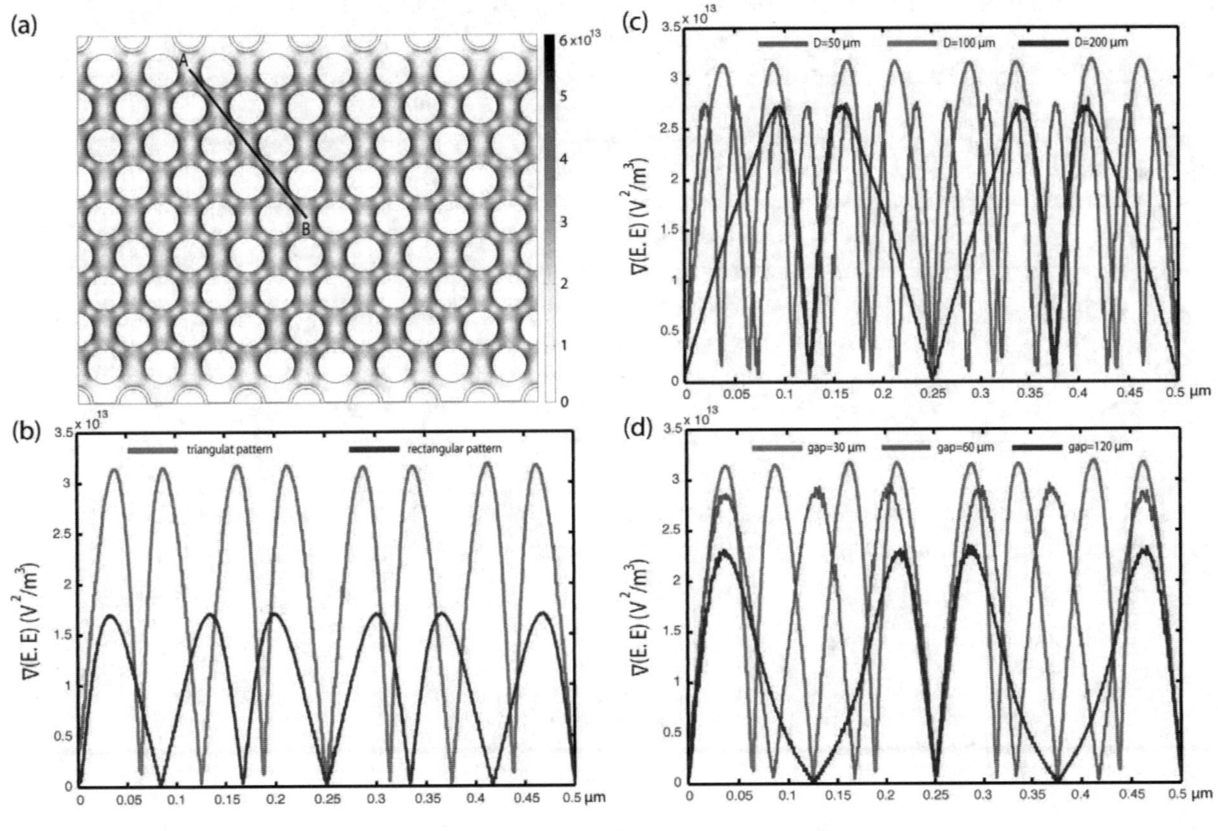

Figure 2 (a) The gradient of the electric field squared (V^2 m^{-3}) at 300 V_{rms} and 600 kHz. The line plot of the gradient of the electric field squared $\nabla(\vec{E}_{rms} \cdot \vec{E}_{rms})$ on a non-dimensionalized diagonal line A-B is presented for (b) triangular and rectangular pillar arranegments, (c) circular pillars with 50, 100, and 200 μm in diameter and 30 μm gap between pillars , (d) circular pillars with 30, 60 and 120 μm gap and diameter of 100 μm, at 100 V_{rms} and 500 kHz.

Fluid dynamics was also modeled to find velocity field and shear rate in the main fluidic channel. No slip boundary conditions were applied to the walls of the channel and the pillars. The inlet velocity was set to 110 μm s^{-1} based on the flow rate in the experiments and the outlet was set to no viscous stress (Dirichlet condition on the pressure). The Navier-Stokes equation was solved for an incompressible laminar flow. The viscosity and density of water, 0.001 Pa.s and 1000 kg m^{-3}, respectively, were used in the main fluidic channel. Computational results predict that the maximum shear rate in the device is approximately two orders of magnitude less than the cell lyses limit.

In order to trap a particle in a moving fluid, the dielectrophoretic force should overcome the drag force. Thus, both drag and dielectrophoretic forces should be considered in designing cDEP microdevices. We investigated the ratio of drag force to dielectrophoretic force for three microdevices with different pillar sizes. Figure 3 presents drag force to dielectrophoretic force ratio for pillars with diameter of 50 μm, 100 μm, and 200 μm. In general, by decreasing the size of pillars, the chance of cell trapping increases, but the trapping area of each pillar, the area in which the dielectrophoretic force dominates drag force (blue areas in Figure 3), decreases. On the other hand, by increasing the size of pillars the area of cell trapping increases but the chance of particle trapping decreases because of fewer number of pillars. Diameters of 100 μm are optimum for cDEP microdevices under these conditions.

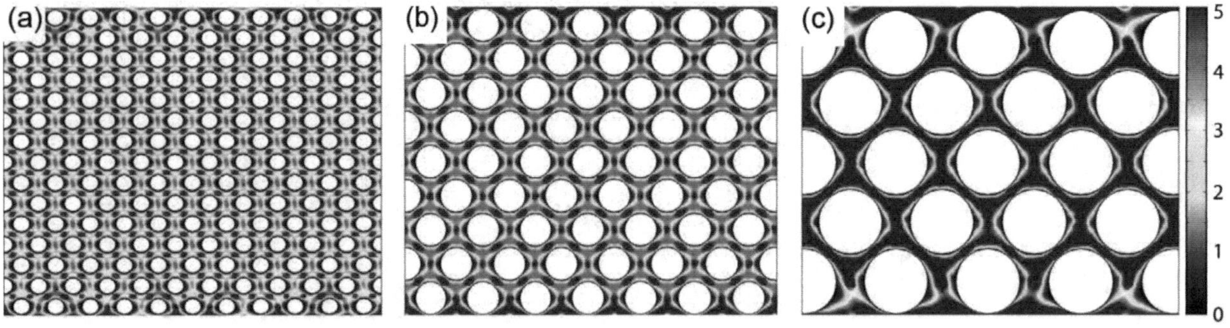

Figure 3. Ration of drag force to dielectrophoretic force for microdevices with equal channel width and circular pillars with diameter of (a) 50 μm, (b) 100 μm, and (c) 200 μm.

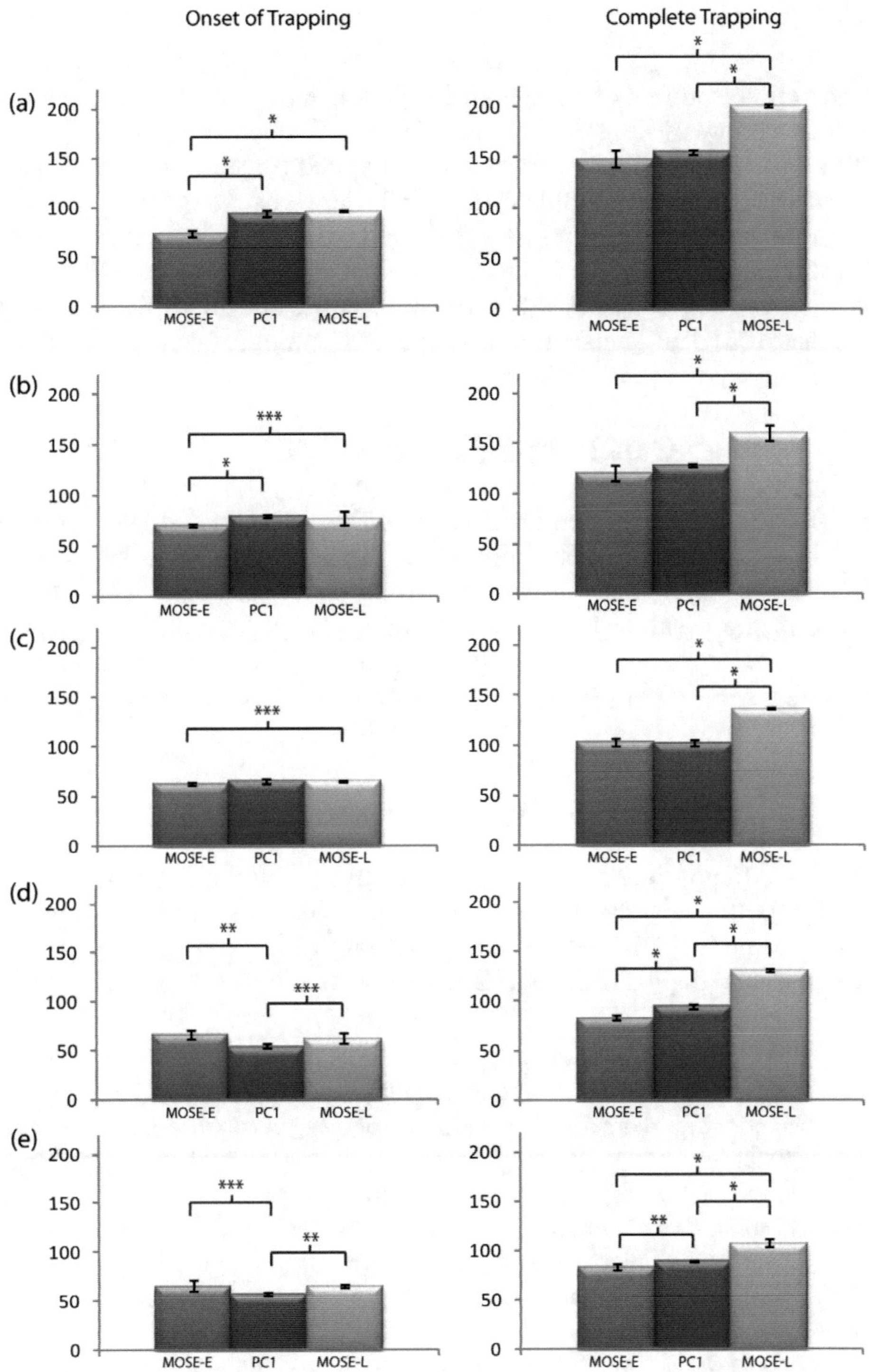

Figure 4. Trapping Voltage of MOSE early and late stages and PC1 cells at (a) 200 kHz, (b) 300 kHz, (c) 400 kHz, (d) 500 kHz, and (e) 600 kHz. n=5 for MOSE early and late cells and n=3 for PC1 cells. Left and right columns present the onset of trapping and complete trapping, respectively. *, **, and *** represent that data are significantly different with $p < 0.0005$, $p < 0.005$, and $p < 0.05$, respectively.

High frequency cDEP

The dielectrophoretic responses of MOSE early, MOSE late, and PC1 cells were studied separately using the device shown in Fig. 1. For five selected frequencies (200, 300, 400, 500, and 600 kHz) the voltage at which at least two cells were trapped (onset of trapping), as well as the voltage at which all cells trapped (complete trapping) were observed for PC1 and MOSE early and late stages. Results are presented in Fig 4 where, for each frequency, it can be observed that MOSE late cells fully trapped at a higher voltage than that of early stage and PC1 cells. PC1 cells also trapped on higher voltage than MOSE early stage cells, but the difference in trapping voltages was less than that of MOSE late stage. The onset of trapping of cells occurred at closer voltages, comparing to complete trapping voltages.

Concluding Remarks

Utilizing the cDEP technique, we demonstrated that MOSE early and late stages, as well as macrophages (PC1) have different dielectric properties. These results are a key step towards enabling cDEP-based microfluidics a technology platform for physicians and biologists to screen for ovarian cancer through the peritoneal fluid. Future microdevices should be able to isolate ovarian cancer cells from background, which will not only helps physicians to diagnose the ovarian cancer in early stages, but to also facilitate further analysis such as genotyping and drug testing.

Acknowledgements

This material is based upon work supported in part by the National Science Foundation under Grant No. EFRI 1107049, by the Virginia Tech Institute for Critical Technology and Applied Science (ICTAS), and the Diversity Summer Research Program (DSRP). The authors would like to thank Chun Cao, Amanda Shea, and BEMS labs members for their contributions.

References

[1] A. Jemal, *et al.*, "Global cancer statistics," *CA Cancer J Clin,* vol. 61, pp. 69-90, 2011.

[2] I. Visintin, *et al.*, "Diagnostic markers for early detection of ovarian cancer," *Clin Cancer Res,* vol. 14, pp. 1065-1072, 2008.

[3] I. J. Jacobs and U. Menon, "Progress and challenges in screening for early detection of ovarian cancer," *Mol Cell Proteomics,* vol. 3, pp. 355-366, 2004.

[4] V. Nossov, *et al.*, "The early detection of ovarian cancer: from traditional methods to proteomics. Can we really do better than serum CA-125? ," *Am J Obstet Gynecology,* vol. 199, pp. 215-223, 2008

[5] R. Pethig, "Review Article-Dielectrophoresis: Status of the theory, technology, and applications " *BIOMICROFLUIDICS,* vol. 4, 2010.

[6] K. Khoshmanesh, *et al.*, "Dielectrophoretic platforms for bio-microfluidic systems " *BIOSENS BIOELECTRON,* vol. 26, pp. 1800-1814, 2011

[7] E. B. Cummings and A. K. Singh, "Dielectrophoresis in Microchips Containing Arrays of Insulating Posts: Theoretical and Experimental Results," *Anal Chem,* vol. 75, pp. 4724-31, 2003.

[8] C. Chou, *et al.*, "Electrodeless Dielectrophoresis of Single- and Double-Stranded DNA," *Biophys J,* vol. 83, pp. 2170-2179, 2002.

[9] Lapizco-Encinas, *et al.*, "Insulator-based dielectrophoresis for the selective concentration and separation of live bacteria in water," *Electrophoresis* vol. 25, pp. 1695–1704, 2004.

[10] R. Gallo-Villanueva, Jesús-Pérez, N., Martínez-López, J., Pacheco, A., Lapizco-Encinas, B., "Microfluid. Nanofluid," vol. 10, pp. 1305–1315, 2011.

[11] H. Shafiee, *et al.*, "Contactless dielectrophoresis: a new technique for cell manipulation," *Biomedical Microdevices,* vol. 11, pp. 997-1006, 2009.

[12] H. Shafiee, *et al.*, "Selective isolation of live/dead cells using contactless dielectrophoresis (cDEP)," *Lab on a Chip,* vol. 10, pp. 438-445.

[13] M. B. Sano, *et al.*, "Contactless Dielectrophoresis Spectroscopy: examination of the dielectric propoerties of cells found in blood," *Electrophoresis,* 2011, in press.

[14] M. B. Sano, *et al.*, "Modeling and development of a low frequency contactless dielectrophoresis (cDEP) platform to sort cancer cells from dilute blood samples," *Bioelectronics and Biosensors,* 2011, in press.

[15] H. A. Pohl, *Dielectrophoresis*: Cambridge University Press, 1978.

[16] M. E. Roberts PC, Baxa AC, "Sequential molecular and cellular events during neoplastic progression: a mouse syngeneic ovarian cancer model," *Neoplasia* vol. 7, pp. 944-56, 2005.

[17] S. W. Creekmore AL, Cimini D, Jensen RV, Roberts PC, Schmelz EM. , "Changes in gene expression and cellular architecture in an ovarian cancer progression model. ," *PLoS ONE,* vol. 6, p. e17676, 2011.

[18] L. A. Flanagan, *et al.*, "Unique dielectric properties distinguish stem cells and their differentiated progeny," *Stem Cells,* vol. 26, pp. 656-65, 2008.

[19] S. Nagrath, *et al.*, "Isolation of rare circulating tumour cells in cancer patients by microchip technology," *Nature,* vol. 450, pp. 1235-1239, 2007.

[20] P. R. C. Gascoyne, *et al.*, "Dielectrophoretic Separation of Cancer Cells from Blood," *Industry Applications, IEEE Transactions on* vol. 33, pp. 670 - 678, 1997.

(276e) Development and Characterization of Degron-Based Substrates Capable of E3 Ligase-Mediated Ubiquitination

Adam Melvin, Marcey Waters and Nancy Allbritton, Chemistry, Univeristy of North Carolina, Chapel Hill, NC

The ubiquitin-proteasome pathway is responsible for the degradation of misfolded, damaged, or tightly regulated proteins. A protein is marked for degradation by the transfer, and subsequent polymerization, of the small molecule ubiquitin. Ubiquitination is governed by a series of enzymatic steps culminating in the recognition of a specific protein by the E3 ubiquitin ligases. An E3 ligase can target a protein through the identification of a specific amino acid degradation sequence, or degron. Improper regulation of the ubiquitin-proteasome pathway has been implicated in the development of cancers such as multiple myeloma and acute myeloid leukemia. In fact, targeting this pathway is one of the leading directions in cancer chemotherapeutics; however there are no clinically available biochemical assays that measure protein ubiquitination and proteasome activity in single primary tumor cells before and after treatment. In this study we sought to design and characterize a series of reporter substrates, based on degrons, capable of intracellular ubiquitination, as a first step in the generation of single cell reporters of E3 ligase and proteasome-targeted therapeutics.

Using solid phase peptide synthesis (SPPS), we developed small (~25-30 amino acids) reporters consisting of four essential components: the E3-ligase specific degron, a proximal lysine residue to be ubiquitinated, a linker to avoid steric hindrance issues between the lysine and the degron, and a fluorescein tag to serve as a visual marker. The degrons were selected across a wide array of cellular functions from protein targets associated with various cancers including β-Catenin, SRC3, iNOS, HIF-1α, TAZ and others. Reporter ubiquitination is being studied in a HeLa S100 cytosolic lysate-based assay, a system consisting of a wide range of E3 ligases and an ideal choice for studying an array of potential targets. The S100-based assay is being optimized for a well studied degron, β-Catenin, by determining the proper assay concentrations and components. We will report the extent of ubiquitination of all reporters using gel and capillary electrophoresis (CE), effectively separating the reporters into 'bins' based on the extent of ubiquitination, calculated based on ubiquitinated reporter relative to parent reporter. Our intent is that these results will provide the groundwork for the development of a ubiquitinatable substrate, which we will then incorporate into a more complex, single-cell reporter of E3 ligase and proteasome-targeted therapeutics.

See more of this Session: Biomedical Diagnostics
See more of this Group/Topical: Topical 3: 2011 Annual Meeting of the American Electrophoresis Society (AES)

(276f) Designing Ion Selective Membranes for Biosensing

Satyajyoti Senapati, Zdenek Slouka and Hsueh-Chia Chang, Chemical & Biomolecular Engineering, University of Notre Dame, Notre Dame, IN

A number of interesting phenomena occur at the interface of an ion selective environment (nanochannel or ion exchange membrane) and the surrounding electrolyte under DC or AC electric fields. These phenomena include formation of depletion and concentration regions around the nanoporous membrane or nanochannel, development of a space charge region with the length exceeding the length of Debye layer, development of electrically driven vortices, Warburg-like Electrical Impedance Spectroscopy (EIS) etc. Our efforts are aimed at using such ion selective environments to develop a highly specific and sensitive DNA/RNA biosensing platform. The biosensing is based on a charge inversion phenomenon that occurs when negatively charged DNA/RNA molecules hybridize with specific probes immobilized on an ion exchange membrane bearing a positive fixed charge. Expensive and bulky optical biosensing technologies can thus be replaced by miniature electronic circuits connected to on-chip nanoporous membrane sensors. Such nanoprobes are not grown or fabricated by expensive nanofabrication techniques, but are rather loaded on-site by patented dielectrophoretic assembly technologies or by standard sol-gel or UV curable chemistry. Ion depletion action by the nanoporous membranes controls the ionic strength around the sensor to ensure robustness to sample variability. The ion-depleted region also dielectrophoretically trap and concentrate the target molecules from the flowing sample to hybridize/dock with molecular probes functionalized onto the nanosensors. Instead of the classical electrochemical sensing technique, we utilize a much more sensitive conductance and impedance signature based on sensitivity of ion current across the nanoporous membrane to surface charge density change effected by hybridized molecules. The sensitivity involves a nonlinear correction to the conductance and is hence much more sensitive than EIS or other linear electrochemical sensing techniques.

See more of this Session: Biomedical Diagnostics
See more of this Group/Topical: Topical 3: 2011 Annual Meeting of the American Electrophoresis Society (AES)

(276g) Dielectrophoretic Separation of Cancer Cells Based On Metastatic Stage

Andrea Diane Rojas, Materials Science and Engineering, Virginia Tech, Blacksburg, VA and Rafael Davalos, School of Biomedical Engineering and Sciences, Virginia Polytechnic Institute and State University, Blacksburg, VA

Cancer is the second leading cause of death in the United States and early detection and classification of stage is critical for the successful treatment of cancer. Several methods have been studied to detect cancerous cells, but compromise cell structure as well as further tests, such as *in vitro* therapeutic treatments, that can improve prognosis. One of the most promising techniques utilized on microfluidic chips for the separation of biological particles is dielectrophoresis (DEP), or the motion of a particle relative to a fluid due to a non-uniform electric field. This phenomenon uses a cell's intrinsic properties to its advantage without altering the original structure or disrupting the viability of the cell. The present work shows progress towards improving prognosis by separating and isolating cancer cells based on their metastatic stage for cultivation. The method, contactless dielectrophoresis, successfully separates breast cancer cells and improves sample purity by eliminating sample/electrode contact which may lead to contaminating effects.

See more of this Session: Biomedical Diagnostics
See more of this Group/Topical: Topical 3: 2011 Annual Meeting of the American Electrophoresis Society (AES)

(320a) Nanoscale Electrokinetics of Correlated Electrolytes and Ionic Liquids

Martin Bazant, Chemical Engineering, MIT, Cambridge, MA and Brian D. Storey, Franklin W. Olin College of Engineering, Needham, MA

Perhaps the most basic assumption of classical electrokinetic theory is the mean-field approximation, where the each ion feels only the electric field produced by the mean charge density (via Poisson's equation) rather than the fluctuating Coulomb forces with individual neighbors. Here, we present a simple Landau-Ginzburg-type continuum model for electrostatic correlations between finite-sized ions [1], which leads to a 4th order modified Poisson equation, convenient for the analysis of electrokinetic phenomena. When the mean-field approximation breaks down, e.g. due to large ion concentrations, large ion valences, and/or nanoscale confinement, the zeta potential loses its significance, and the model predicts that electro-osmotic flows are typically reduced – or even reversed - by correlation effects, compared to the prediction of the Helmholtz-Smoluchowski formula. This may help to explain the over-prediction of induced-charge electro-osmotic flows by classical models [2]. An interesting limit of the model describes electro-osmosis in solvent-free ionic liquids and molten salts, which may be important in energy storage and electroactuation applications.

[1] M. Z. Bazant, B. D. Storey, and A. A. Kornyshev, Phys. Rev. Lett. **106**, 046102 (2011).

[2] M. Z. Bazant, M S Kilic, B Storey, and A Ajdari, Adv. Colloid Interface Sci., 152, 48-88 (2009).

See more of this Session: Nanoscale Electrokinetics
See more of this Group/Topical: Topical 3: 2011 Annual Meeting of the American Electrophoresis Society (AES)

(320b) Efficiently Accounting for Ion Correlations In Electrokinetic Nanofluidic Devices Using Density Functional Theory

Dirk Gillespie, Molecular Biophysics and Physiology, Rush University Medical Center, Chicago, IL and Sumita Pennathur, Mechanical Engineering, University of California at Santa Barbara, Santa Barbara, CA

The electrokinetic behavior of nanofluidic devices is dominated by the electrical double layers at the device walls. Therefore, accurate, predictive models of double layers are essential for device design and optimization. We demonstrate that density functional theory (DFT) of electrolytes is an accurate and computationally efficient method for computing finite ion size effects and the resulting ion-ion correlations that are neglected in classical double layer theories such as Poisson-Boltzmann. DFT is designed for nanofluidic systems with small spatial dimensions, high surface charge densities, high ion concentrations, and/or large ions. These regimes produce nonlinear phenomena such as charge inversion, wherein more counterions adsorb at the wall than is necessary to neutralize its surface charge, leading to a second layer of co-ions. We show that DFT can predict charge inversion and other nonlinear phenomena. DFT reproduces experimental results of ion current in single-slit nanofluidic devices and gives insights into the limits of energy conversion via streaming currents and molecular sensing.

See more of this Session: Nanoscale Electrokinetics
See more of this Group/Topical: Topical 3: 2011 Annual Meeting of the American Electrophoresis Society (AES)

(320c) Highly Efficient Electrophoretic and Electroosmotic Flow Through Carbon Nanotube Membranes

Ji Wu[1], Karen Gerstandt[1] and **Bruce J. Hinds**[2], (1)Chem.&Mater Engr., University of Kentucky, Lexington KY, KY, (2)Chemical and Materials Eng., University of Kentucky, Lexington KY, KY

Carbon nanotubes have three key attributes that make them of great interest for novel membrane applications 1) atomically flat graphite surface allows for ideal fluid slip boundary conditions and extremely fast flow rates 2) the cutting process to open CNTs inherently places functional chemistry at CNT core entrance for chemical selectivity and 3) CNT are electrically conductive allowing for electrochemical reactions and application of electric fields gradients at CNT tips. The CNT membrane, with tips functionalized with charged molecules, is a nearly ideal platform to induce electro-osmotic flow with high charge density at pore entrance and a nearly frictionless surface for the propagation of plug flow. Through diazonium electrochemical modification we have successfully bound anionic surface charge to CNT tips and along CNT cores. Both cationic and neutral fluxes through the CNT membrane are observed as a function of bias. Ionic mobilities are enhanced by a factor of ~5-8 with rectification seen for large anion/cation mixtures proving conducting through CNTs. High electro-osmotic flows of 0.05 cm/s at negative 300mV bias are seen. The analysis of charged dye molecules (Ru (bpy)3^2+) showed a factor of 50 enhancement in on state (-300mV) and of state (+300mV) primarily due to electrophoretic effects. Neutral molecules (caffeine) within ionic solutions showed a factor of 10 enhancement in on state (-300mV) and of state (+300mV). The enhanced transport of neutral species under bias is the primary proof of the existence of an electro-osmotic processes. The optimization of this process occurred after 2 steps; 1) use of CNTs with diameter under 2nm to avoid the effects of screening (Debye length ~1nm) and 2) use of quad-valent charged dye tethered to CNT entrance and core. Use of the electro-osmotic phenomenon for responsive/programmed transdermal drug delivery devices is discussed with the voltage gated delivery of clonidine and nicotine across CNT membrane at therapeutically useful fluxes.

[1] "Aligned Multiwalled Carbon Nanotube Membrane" Hinds, B.J.*; Chopra, N.; Andrews, R.; Gavalas, V.; Bachas L. Science 2004 303 62-65.

[2] "Nanoscale hydrodynamics: Enhanced flow in carbon nanotubes" Majumder, M.; Chopra, N.; Andrews, R; Hinds, B.J * Nature 2005, 438, 44.

[3] "Voltage Gated Carbon Nanotube Membranes" Majumder, M.; Zhan, X; Andrews, R; Hinds, B.J * Langmuir 2007; 23(16); 8624-8631.

[4] " Programmable transdermal drug delivery of nicotine using carbon nanotube membranes" J. Wu, K. S. Paudel, C.L. Strasinger, D. Hamell, Audra L. Stinchcomb*, B. J. Hinds* *Proc. Nat. Acad. Sci.* **2010** 107(26) 11698

See more of this Session: Nanoscale Electrokinetics
See more of this Group/Topical: Topical 3: 2011 Annual Meeting of the American Electrophoresis Society (AES)

(320d) Rectification, Hysterisis and Oscillations In Nanoscale Electrokinetics

Hsueh - Chia Chang[1], Li-Jing Cheng[2], Zdenek Slouka[2], Yu Yan[2] and Anees Attarwala[2], (1) Department of Chemical and Biomolecular Engineering, University of Notre Dame, Notre Dame, IN, (2) Chemical and Biomolecular Engineering, University of Notre Dame, Notre Dame, IN

Our group has been studying several nonlinear ion current phenomena in ion-selective nanopores, nanoslots and nanoporous membranes under an applied DC field. We are able to design nanofluidic components that exhibit rectifying and excitable/oscillatory ion-current dynamics of biological membranes with protein ion channels. We are designing devices with such nonlinear current/voltage features to improve the performance of biosensors and mass spectrometry.We have shown that the ion current rectification phenomenon first observed by Z. Siwy (Nature Nanotechnology, 2008) when a low voltage is applied across a conic pore is due to internal and external depletion/enrichment driven by asymmetric conductivity jumps at the two ends. A cross-section averaged theory quantitatively captures the rectification factor and is favorably compared to simulations and experimental measurements. Beyond a critical voltage related to the Donnan potential of the pore, the rectification is due to asymmetric external depletion at the two ends (G. Yossifon and H-C Chang, Phys Rev Lett, 2009) and is opposite in direction as low-voltage rectification (L-J Chen and L-J Guo, Chem Soc Rev, 2010). For symmetric electrolyte, the concentration profile in the electroneutral depletion zone is governed by the diffusion Laplace equation and the classical solution for a disk sink provides an estimate of the large-voltage rectification factor. We report the first experimental data on this rectification reversal phenomenon at a critical voltage that is consistent with our theory. Generalization to other nanofluidic diodes with different spatial asymmetries will also be discussed. We have also experimentally and theoretically examined the use of external depletion to concentrate analytes, as was first reported by J. Han's group (Anal Chem, 2005). It is found that the concentration only occurs for minority coion analytes. The concentration slug develops at a particular location near the boundary of the depletion zone due to a balance of electrophoretic flux and convective flux. Analyte stacking is possible for analytes of different mobilities but vortices produced by pressure-driven back flow, to ensure flow continuity, can destroy the concentrated slugs. With proper channel/membrane design, based on our theoretical analysis, we are able to produce negative differential resistance, excitability, bistability and hysteresis. Upon coupling this nonlinear ion circuit with a linear electronic capacitor, the hybrid ion-electron nonlinear circuit can sustain relaxation type oscillations in the voltage and current. This oscillatory nanofluidic clock is used to concentrate and release specific analytes periodically for an on-chip biosensor and for a continuous spray emitter for mass spectrometry.

An exciting new direction is to use feedback control to sustain the local pH, ionic strength and electric field to produce non-equilibrium chemical kinetics in nanoscale electrokinetics. Two companion papers on water-breaking reactions for pH actuation and DNA hybridization detection will be presented by L.-J. Chen and Z. Slouka.

(320e) Micro- pH Tuner and Its Applications Based On Field-Enhanced Water Dissociation In Bipolar Nanoporous Membranes

Li-Jing Cheng, Chemical & Biomolecular Engineering, University of Notre Dame, Notre Dame, IN and H.-C. Chang, Chemical and Biomolecular Engineering, University of Notre Dame, Notre Dame, IN

Control of pH in microscale environments offers many intriguing possibilities for several biological applications such as on-chip microbioreaction (proteolysis, protein crystallization etc) and cell or biomolecular separation. Conventional methods for changing local pH include diffusion of CO_2, addition of acid and base solutions and electrolysis. These methods, however, are either difficult to miniaturize or require microvalves to control the flow of different pH solutions. Electrolysis produces hydrogen and oxygen bubbles that impede microfluidic flow and other reaction products that contaminate the sample. To overcome the aforementioned obstacles, we present a novel strategy to regulate pH in in microscale environments by injecting controllable amounts of protons and hydroxide ions via field-enhanced dissociation of water molecules. The dissociation occurs at the ion-depleted junction of reverse-biased bipolar membranes, fabricated by photo-patterning ion-exchange polymers into microfluidic chips. The bipolar membrane generates and separates H^+ and OH^- ions without gas production or contaminant generation of electron-transfer reactions. Furthermore, it effects pH changes with no need of additional acidic/basic solutions that dilute analyte concentrations. Combining with properly designed microchannel configurations, the approach allows us to sustain robust local on-chip pH and pH gradients to improve several microfluidic operations, such as micro-reactors and molecular separation in microfluidic isoelectric focusing chips.

See more of this Session: Nanoscale Electrokinetics
See more of this Group/Topical: Topical 3: 2011 Annual Meeting of the American Electrophoresis Society (AES)

(320f) Measurement of the Zeta Potential and Darcy Coefficient of Porous Materials by Rotating Samples On Their Axis

Paul Sides, Shravya Mukka and Dennis C. Prieve, Chemical Engineering, Carnegie Mellon University, Pittsburgh, PA

A relatively new method of zeta potential measurement utilizes a planar disk sample affixed to the end of a spindle and rotated. [1] The theory underlying the method is based on the well established hydrodynamics of the rotating disk. The simplicity and convenience of the apparatus, however, invites users to affix samples of various sizes, shapes, and structures to the spindle and obtain a streaming potential; however, one must convert the measured streaming potential to zeta potential with a theory suitable for the particular geometry of the sample. Here, porous samples having a finite thickness and open structure were placed on the spindle and rotated. Large streaming potentials were generated and naïve application of the theory developed for planar surfaces gave proportionally large apparent zeta potentials. A treatment of the fluid flow in a rotating porous sample was then found in the fluid mechanics literature [2] and applied to this problem. An analytical result relating the streaming potential to the zeta potential of the porous material, its Darcy law coefficient, and the thickness of the sample, was obtained. The streaming potential depends on the square of the rotation rate at low rates and becomes independent of rotation rate at high rates. Experiments combined with the new theory, both of which are described in this presentation, have demonstrated that one can determine the zeta potential and the permeability of porous samples by means of this approach. Plotting the reciprocal of the measured streaming potential versus the reciprocal of the square of the rotation rate gives a straight line having a slope independent of the porous structure of the sample. A plot of the quotient of the streaming potential and the square of the rotation rate against the streaming potential gives a straight line with a slope that only depends of the Darcy coefficient of the porous sample.

[1] 1. P. Sides, J. Newman, J. Hoggard, D. Prieve, *Langmuir* **22** 9765-9769 (2006).

[2] D. D. Joseph, *Q. J. Mech. Appl. Math.* **18** 325 (1965)

See more of this Session: Nanoscale Electrokinetics
See more of this Group/Topical: Topical 3: 2011 Annual Meeting of the American Electrophoresis Society (AES)

(320g) Symmetry Breaking During AC Electrophoresis Normal to An Electrode

Christopher L. Wirth, Paul J. Sides and Dennis C. Prieve, Department of Chemical Engineering, Carnegie Mellon University, Pittsburgh, PA

AC electrophoresis of a single particle proximate to an electrode is important to the 2D crystallization of colloidal particles.[1] Particle assembly is a complicated function of electric field strength and frequency. Fagan et. al.[2] observed that a break in symmetry in a single particle's motion along the direction of the applied electric field strongly correlated with the lateral equilibrium spacing of a 2D colloidal crystal. The break in symmetry was captured in the phase angle (θ) between the particle's height above the electrode and the electric field driving motion. To date, no *a priori* model can predict this phase angle. Here, we present a model that predicts a symmetry-breaking phase angle for the first time. The full model, including colloidal and hydrodynamic forces, and including the Basset force not previously cited in this context, showed that θ can increase from 0° at low frequencies, cross 90° at ~100 Hz and then increase to 180° as frequency was increased. The remaining unknown is the mechanism underlying the role of ion identity in particle motion. Particles repel each other in KOH during directed assembly but attract each other in $NaHCO_3$. We now believe that this electrolyte dependence arises in the apparent electrokinetic charge on the particle, which can be approximated by the Stokes drag coefficient multiplied by the dynamic electrophoretic mobility of the particles. The latter quantity is known to depend on the ion mobilities. As a first inquiry, we calculated the frequency dependent electrophoretic mobility of the particle as a function of the dispersing electrolyte's ion mobilities. Preliminary results from this updated model suggest the particle's dynamic electrophoretic mobility is important to solving the mystery of electrolyte dependent particle motion.

1. Prieve, D. C.; Sides, P. J.; Wirth, C. L., 2-D assembly of colloidal particles on a planar electrode. Current Opinion in Colloid & Interface Science (2010) *15* (3).

2. Fagan, J. A., Sides, P. J.; Prieve, P. C., Vertical oscillatory motion of a single colloidal particle adjacent to an electrode in an ac electric field. Langmuir (2002) *18* (21).

See more of this Session: Nanoscale Electrokinetics
See more of this Group/Topical: Topical 3: 2011 Annual Meeting of the American Electrophoresis Society (AES)

(320h) Low-Frequency Dielectrophoretic Response of a Single Particle In Aqueous Suspensions

Jingyu Wang and H. Daniel Ou-Yang, Physics, Lehigh University, Bethlehem, PA

Alpha-relaxation, the counterion diffusion in the electric double layer, has been used to describe the anomalous low frequency dielectric dispersion of aqueous suspensions of colloidal particles. A microscopic theory describing this relaxation process proposed by Schwarz, however, has not been investigated systematically. We use an optical tweezers-based single-particle dielectrophoretic (DEP) force spectroscopy to study the relaxation mechanism as a function of particles size, temperature and solvent viscosity. Specifically, we measure the dependence of the DEP crossover frequency and compare results with predictions by Schwarz.

See more of this Session: Nanoscale Electrokinetics
See more of this Group/Topical: Topical 3: 2011 Annual Meeting of the American Electrophoresis Society (AES)

(320j) Streaming Potential Revisited

Ehud Yariv[1], Ory Schnitzer[1] and Itzchak Frankel[2], (1)Mathematics, Technion - Israel Institute of Technology, Haifa, Israel, (2)Aerospace Engineering, Technion - Israel Institute of Technology, Haifa, Israel

Electrokinetic streaming-potential phenomena are driven by imposed relative motion between liquid electrolytes and charged solids. Owing to non-uniform convective `surface' current within the Debye layer Ohmic currents from the electro-neutral bulk are required to ensure charge conservation thereby inducing a bulk electric field. This, in turn, results in electro-viscous drag enhancement. The appropriate modeling of these phenomena in the limit of thin Debye layers has been a matter of ongoing controversy apparently settled by Cox's seminal analysis (J. Fluid Mech., vol. 338, 1997, p. 1). This analysis predicts electro-viscous forces resulting from the perturbation of the original Stokes flow with the Maxwell-stress contribution only appearing at higher orders.

Cox's theory is founded upon the assumption of O(1) Hartmann and Peclet numbers. We demonstrate that the product of these numbers actually scales inversely with the square of Debye width and accordingly revisit the generic problem of streaming-potential. Electric-current matching between the Debye layer and the bulk provides an inhomogeneous Neumann condition governing the electric field in the latter. This field, in turn, results in a velocity perturbation generated by a Smoluchowski-type slip condition. Owing to dominant convection, the present analysis yields an asymptotic structure considerably simpler than that of Cox: the electro-viscous effect now appears at a lower asymptotic order and is contributed by both Maxwell and viscous stresses. The present paradigm is illustrated for the prototypic problem of a sphere sedimenting in an unbounded fluid domain with the resulting drag correction differing from that calculated by Cox.

Independently of current matching, salt-flux matching between the Debye layer and the bulk domain need also be satisfied. This subtle point has apparently gone unnoticed in the literature, perhaps because it is trivially satisfied in field-driven problems. In the present limit this requirement seems incompatible with the uniform salt distribution in the convection dominated bulk domain. This paradox is resolved by identifying the dual singularity associated with the thin-Debye-layer limit in motion-driven problems resulting in an intermediate diffusive layer beyond the familiar Debye layer.

See more of this Session: Nanoscale Electrokinetics
See more of this Group/Topical: Topical 3: 2011 Annual Meeting of the American Electrophoresis Society (AES)

(320i) Electrokinetic Lift In Shear Flows

Ory Schnitzer[1], Ehud Yariv[1] and Itzchak Frankel[2], (1)Mathematics, Technion, Haifa, Israel, (2) Aerospace Engineering, Technion, Haifa, Israel

Electrolyte flow relative to a charged surface induces a bulk electric field (the ``streaming potential'' phenomena). This field, and the flow perturbation it animates, generate both electrical and hydrodynamic ``electro-viscous'' forces whose magnitude has been a matter of ongoing controversy. Recently, we have revisited this problem, showing that these forces scale as the second power of Debye width, as opposed to earlier predictions of fourth and sixth powers.

Electro-viscous forces could explain several surprising observations involving the motion of colloidal particles at low-Reynolds-number conditions, such as the anomalous repulsion of microspheres from an adjacent wall in the presence of an imposed shear flow, as observed by Prieve and co-workers. Owing to the symmetry properties of the linear Stokes equations, such repulsion is inadmissible in the absence of inertial effects. These, in turn, have been ruled out in the experiments.

This particle--wall interaction is analyzed using our revised model. The undisturbed flow consists of three components: the `driving' mechanism, a stationary sphere in a shear flow, together with the `induced' components, particle translation parallel to the wall and rotation normal to it, which follow from the constraint of a force-free particle. Symmetry arguments show that the additional electro-viscous lift results from Maxwell stresses, while the additional drag results from viscous stresses.

Using a lubrication approximation we consider the case where the dimensionless particle--wall gap is small. At leading-order, both the lift and additional drag are contributed by the gap region, scaling as negative powers of its width. It is noteworthy that the streaming-potential mechanism underlying these forces arises from the `induced' flow components, rather than the driving shear component. The contributions associated with translation and rotation are comparable. For equal zeta potentials, electro-viscous drag are actually proportional to the sum of translational and rotational velocities, while electro-viscous lift is proportional to its square.

See more of this Session: Nanoscale Electrokinetics
See more of this Group/Topical: Topical 3: 2011 Annual Meeting of the American Electrophoresis Society (AES)

(365b) Rapid, Gel-Free Electrophoretic Separation of DNA Oligonucleotides Using Surfactant Systems: Extended Read Frame by Buffer Design

Stephen Istivan, Angela Holmen, Max A. Fahrenkopf and **James W. Schneider**, Chemical Engineering, Carnegie Mellon University, Pittsburgh, PA

We present some significant improvements to the "micelle-ELFSE" method we have developed for the rapid electrophoretic separation of end-alkylated DNA oligonucleotides (aDNA) in the presence of nonionic surfactant micelles. In capillary electrophoresis (CE) format, the aDNA are injected into the capillary pre-filled with a running buffer containing nonionic surfactant above its critical micelle concentration. On application of electric field, the aDNA migrate through the capillary and attach to the micelles, retarding their mobility in a length-dependent way. The rapid statistical sampling of micelle aggregation number ensures that a highly uniform retardation is experienced by all aDNA in the sample, yielding very sharp peaks under appropriate conditions. As a result, DNA differing by a single base, up to 420 bases in length, can be distinguished in less than 15 minutes using commercial CE equipment. This represents an 8-10 fold decrease in runtime compared to commercial capillary gel electrophoresis methods.

Here, we present a theory that predicts the best possible separation schemes given the timescale of micelle size fluctuations, the polydispersity of the micelles, diffusion-based broadening, injection and detection-based broadening, and wall adsorption. Interestingly, we find that minimum run times required for each possible DNA read frame are set only by the polydispersity and fluctuation dynamics of micelles in the running buffer, not by the capillary length or electric fields used or even the size of the micelles. This insight is crucial for the design of optimal run conditions and buffers for various bioanalysis applications. We demonstrate that, with proper design, the micelle-ELFSE method can be implemented to separate DNA lengths well above the biased reptation limit with single-base resolution. We present some recent applications of the micelle-ELFSE method, including identification of STR loci, miRNA profiling, and single-base-extension sequencing.

See more of this Session: DNA Analysis In Microfluidic and Nanofluidic Devices
See more of this Group/Topical: Topical 3: 2011 Annual Meeting of the American Electrophoresis Society (AES)

(365c) DNA Electrophoresis In a Nanofence Array

Sung-Gyu Park and Kevin D. Dorfman, Chemical Engineering and Materials Science, University of Minnesota, Minneapolis, MN

We will present separation data obtained in a new post array geometry for long DNA electrophoresis that we call the "nanofence" array. Previous post array devices used relatively dense arrays of posts, with typical pore spaces of 0.5 to 5 microns. The nanofence device consists of periodic regions of closely spaced posts (600 nm diameter and 600 nm gap size, arranged in two offset rows), which are the fences. They are separated by a relatively long distance of 20 microns to allow relaxation of the DNA between fences. The idea is to force the DNA to collide with a nanopost in order to pass through the fence. The distance between the collisions then becomes deterministic, and the band broadening only arises due to variations in the holdup time at a fence. We fabricated the device using a combination of projection lithography and deep reactive ion etching. To demonstrate the principle of the separation, we used λ-DNA and its XhoI digest. These species were baseline resolved after a few minutes with a significant reduction in band broadening.

See more of this Session: DNA Analysis In Microfluidic and Nanofluidic Devices
See more of this Group/Topical: Topical 3: 2011 Annual Meeting of the American Electrophoresis Society (AES)

(365d) Saturation of the Cross-Stream Migration of DNA Driven by a Pressure Gradient and An Electric Field

Jason Butler[1], Rahul Kekre[2], Julian Hook[3] and A.J.C Ladd[1], (1)Chemical Engineering, University of Florida, Gainesville, FL, (2)Department of Chemical Engineering, University of Florida, Gainesville, FL, (3)University of Florida, Gainesville, FL

DNA exhibits a transverse migration across streamlines under simultaneous electrophoresis and pressure-driven flow. We have proposed that the migration is driven by long-range hydrodynamic interactions generated by the electric field acting upon the polyelectrolyte, which is distorted from its equilibrium conformation by the local shear. A coarse-grained simulation that includes the hydrodynamic interactions from the imposed flow and electric fields captures the observed migration of DNA molecules. This model also predicts that the extent of migration saturates at a high field strength when holding the flow rate constant. Measurements of DNA in a microfluidic channel confirm that this saturation, previously unobserved, does exist. This finding provides additional evidence in support of the model and the concept that hydrodynamic interactions generated by an electric field acting on a polyelectrolyte are not fully screened.

See more of this Session: DNA Analysis In Microfluidic and Nanofluidic Devices
See more of this Group/Topical: Topical 3: 2011 Annual Meeting of the American Electrophoresis Society (AES)

(365e) Concentration of DNA Using 3D Carbon-Electrode Dielectrophoresis

Rodrigo Martinez-Duarte[1], Philippe Renaud[1] and Alexandra Ros[2], (1)Microsystems Laboratory, Ecole Polytechnique Fdrale de Lausanne, Lausanne, Switzerland, (2)Chemistry and Biochemistry, Arizona State University, Tempe, AZ

We present the concentration of λ-DNA using volumetric carbon electrodes. Glass-like carbon electrodes featuring a height of 100 um are obtained after the pyrolysis of photopatterned SU-8 structures. A frequency range from 1 kHz to 1 MHz is initially probed. The concentration of DNA is measured as a function of the fluorescence intensity across the electrode array. Initial results show the concentration of DNA in well defined regions in times as short as 20 s. DNA is observed to migrate to regions of high electric field gradient at all the frequencies tested. A higher degree of concentration is observed at the lower end of the frequency range tested. This work is the initial proof of concept towards a high throughput concentrator for DNA analysis. This stage will be integrated to other functions already demonstrated using carbon electrodes such as cell trapping and cell lysis. The goal is to develop a robust and low cost microfluidics device for the extraction and analysis of genetic material from targeted cells.

See more of this Session: DNA Analysis In Microfluidic and Nanofluidic Devices
See more of this Group/Topical: Topical 3: 2011 Annual Meeting of the American Electrophoresis Society (AES)

(365f) Insulator-Based Dielectrophoretic (iDEP) Manipulation of DNA Origami In a Microfluidic System

Lin Gan[1], Baoquan Ding[2], Hao Yan[2] and Alexandra Ros[1], (1)Department of Chemistry and Biochemistry, Arizona State University, Tempe, AZ, (2)Biodesign Institute, Arizona State University, Tempe, AZ

Dielectrophoresis is the movement of particles caused by polarization in a nonuniform electric field. This technique has been used for manipulating, separating and detecting particles using devices within micrometer to submicrometer range. It has potential in coupling with microfluidic systems to build lab-on-a-chip devices for sorting, separation and analysis. Although the method is well established for micrometer-sized particles and biological cells, the underlying mechanism for biomolecules is still little understood. Here, we focus on a fundamental study of DNA dielectrophoresis of structure-defined DNA nanoparticles in an iDEP microfluidic system. DNA was chosen as model biomolecule because it's structure has been well studied and molecular assemblies with defined shapes such as 2D and 3D origamis can be created based on the principal of complementary base pairing. In particular, we explore the polarizability dependent dielectrophoretic behavior of 6-helix bundle and triangle DNA origamis with large structural difference and demonstrate distinct iDEP behavior.

A microdevice to study the iDEP behavior is built of a PDMS channel integrating an array of insulating posts to create electric field gradients upon application of external potential. The channel is 100 μ m wide, 1 cm in length, integrating elliptic posts with a feature size of 6 μ m and 10 μ m along their minor and major axis, respectively. The depth of the channel and the height of the posts are 5 μ m. Numerical simulation with COMSOL is used to determine the electric field distribution in the post array. Within our experimental limits, we thus find the device could achieve 10^5 V/m in electric field and 10^{16} V^2/m^3 in gradient of electric field square (∇E^2). To study the DNA origami's trapping behavior, the DNA molecules are intercalated with a fluorescent dye and DEP manipulation is recorded via fluorescence video microscopy.

The 6-helix bundle origami is a tube of 380 nm length and 7 nm diameter, while the triangle origami is a flat structure with a length of 130 nm along it's symmetric axis and with 2 nm in height. It is expected that 6-helix bundle origami perform a different trapping behavior from triangle origami according to differences in their molecular structure. In our experiment, the sample of around 200 pM concentration is filled in the channel and an electric field of 100 V/cm is applied along the channel. For 6-helix bundles, trapping of the particles around the posts is monitored at different AC frequency. For triangle origami, no trapping is observed, however concentrated streamlines are detected along the posts parallel to the direction of the electric field. We attribute the latter finding to the smaller size and thus reduced polarizability of the DNA triangle origami, resulting in an overall reduced DEP trapping force.

Concentration streamlines arise due to a small pressure driven flow component.

Furthermore, a frequency dependence of the trapping behavior is monitored for the 6-helix bundle origami. As the AC frequency increases, the trapping areas around the posts shrink, while the fluorescence intensity of the trapping area increases. The migration of the DNA particles within DEP traps between the maxima of trapping forces around the insulator posts is also observed. This is most likely caused by the electrophoretic migration within the DEP trap. As the frequency increases, migration due to the electrophoresis is restricted in the trapping region, causing the shrinkage of the trapping area.

In conclusion, we demonstrated the DEP manipulation for DNA origami with the size down to 130nm. The two investigated DNA origamis (6-helix bundle and triangle origami) showed distinct difference in trapping strength. For 6-helix bundle in particular, frequency dependence in the trapping behavior is also observed. Our aim is to exploit the DNA origami trapping behavior quantitatively in the future to reveal the mechanism involved in DNA polarizability and apply the gained knowledge for the optimization of analytical techniques based on DEP.

See more of this Session: DNA Analysis In Microfluidic and Nanofluidic Devices
See more of this Group/Topical: Topical 3: 2011 Annual Meeting of the American Electrophoresis Society (AES)

(365g) Study of DNA Dynamics In Micro/ Nanofluidic De-Wetting Free Surface Flows and Pulsatory Electric Fields

Wei-Ching Liao, Mechanical and Aerospace Engineering, The Ohio State University, Columbus, OH, Orin L. Hemminger, Chemical and Biomolecular Engineering, Ohio State University, Columbus, OH and L. James Lee, Chemical and Biomolecular, the Ohio State University, Columbus

A hybrid micro/nanochannel fluidic device was recently developed in our laboratory to achieve dosage controlled delivery of DNA to single cell using electric pulse induced electrophoresis through the nanochannel, i.e. nanoelectroporation (NEP). The fabrication of circular nanochannel embedded between two microchannels is realized by a DNA combing and imprinting (DCI) technique where large DNA molecules are stretched across a microridge array to form DNA nanowires in a de-wetting process. A molecular imprinting process is then carried out to convert the DNA nanowires to naochannels and microridges to microchannels. In NEP, DNA electrophoresis dynamics in pulsatory electric fields determines the transport of DNA from one microchannel to the other microchannel through the connecting nanochannel. In this work, Brownian dynamics based finite element simulation and nanofluidic experiments are carried out to study both hydrodynamic and electrophoretic DNA dynamics in the aforementioned DNA combing and NEP processes. The detailed DNA dynamics obtained from the Brownian dynamics simulation can explain the experimentally observed DNA nanowire formation and DNA electrophoresis under electric pulses.

See more of this Session: DNA Analysis In Microfluidic and Nanofluidic Devices
See more of this Group/Topical: Topical 3: 2011 Annual Meeting of the American Electrophoresis Society (AES)

(365h) Entropophoresis of a Polymer Chain Confined In a Nanofluidic Staircase

Frederick R. Phelan Jr.[1], Christopher Forrey[2], Samuel M. Stavis[3], Jon Geist[3] and Elizabeth A. Strychalski[4], (1)Polymers Division, NIST, Gaithersburg, MD, (2)Center for Devices and Radiological Health, FDA, Silver Spring, MD, (3)Semiconductor Electronics Division, NIST, Gaithersburg, MD, (4)Biochemical Sciences Division, NIST, Gaithersburg, MD

Entropophoresis ofa Polymer Chain Confined in a Nanofluidic Staircase[*]

FrederickR. Phelan Jr.[][1], Christopher Forrey[2], Samuel M. Stavis[3], JonGeist[3], and Elizabeth A. Strychalski[4]

[1]PolymersDivision, NIST, Gaithersburg, MD 20899

[2]Centerfor Devices and Radiological Health, FDA, Silver Spring, MD 20993

[3]SemiconductorElectronics Division, NIST, Gaithersburg, MD 20899

[4]BiochemicalSciences Division, Division, NIST, Gaithersburg, MD 20899

Abstract

Mass transfer by diffusion can be alternatively thought ofas net molecular transport driven along an entropy gradient, known asentropophoresis. Entropophoresis has recently been observed experimentally as atransport mechanism for single chain-phageDNA confined in a nanofluidic staircase [1-5], a device which consists of a collection of nanoslits of increasing depth arranged in step-like fashion along the length of a channel. A DNA molecule placed in the highly confined region of the geometry and allowed to freely diffuse exhibits a net 1-D drift velocity down the staircase direction, using only changes in entropy due to confinement (i.e., not external forces) to generate the motion.

In this work, we examine the diffusive motion of acoarse-grained polymer chain confined in the nanofluidic staircase using themolecular dynamics simulation software LAMMPS [6-7] for the regime, where is the local gap thickness quantifying the degree of confinement at each stepposition (i.e., step depth), is the Kuhn length, and is the polymer radius of gyration. Simulation results show that a net 1-D driftvelocity is spontaneously generated under conditions where the differential changein entropic free energy of the chain from step to step [8] is greater than the thermal energy available for diffusion. The steps thus effectively function as a Brownian ratchet by retarding the diffusive motion of the polymer in the direction of greater confinement. The drift velocity down the staircasedirection is independent of step height at high confinement, but slows as the localgap thickness approaches.The effect of chain stiffness and the polymer contour length/step size ratioare also considered.

References

[1] E. A. Strychalski, S. M. Stavis, M. Gaitan, andL. E. Locascio, Nanoslinky: DNA Entropophoresis Down A Nanofluidic Staircase,presented at the The 14th International Conference on Miniaturized Systems forChemistry and Life Sciences, Groningen, The Netherlands, 2010, pp. 2071-2073.

[2] S. M. Stavis, J. Geist, M. Gaitan, L. E.Locascio, and E. A. Strychalski, DNA molecules descending a nanofluidicstaircase by entropophoresis, *Submitted*, 2011.

[3] E. A. Strychalski, J. Geist, M. Gaitan, L. E.Locascio, and S. M. Stavis, Multiplexed size measurements of confined DNAmolecules by entropophoresis down a nanofluidic staircase, *Submitted*,2011.

[4] S. M. Stavis, E. A. Strychalski, and M. Gaitan,Nanofluidic structures with complex three-dimensional surfaces, *Nanotechnology*,vol. 20, no. 16, p. 165302, Apr. 2009.

[5] S. M. Stavis, J. Geist, and M. Gaitan,Separation and metrology of nanoparticles by nanofluidic size exclusion, *Labon a Chip*, 2010.

[6] S. Plimpton, Fast Parallel Algorithms forShort-Range Molecular Dynamics, *Journal of Computational Physics*, vol.117, no. 1, pp. 1-19, Mar. 1995.

[7] LAMMPS Molecular Dynamics Simulator. [Online].Available: http://lammps.sandia.gov/. [Accessed: 02-May-2011].

[8] P. G. de Gennes, *Scaling Concepts in PolymerPhysics*. Ithaca: Cornell University Press, 1979.

[]Corresponding author: NIST, 100 Bureau Dr., STOP 8542, Gaithersburg, MD 20899; (301) 975-6761; Fax: (301) 975-4924; frederick.phelan@nist.gov

See more of this Session: DNA Analysis In Microfluidic and Nanofluidic Devices
See more of this Group/Topical: Topical 3: 2011 Annual Meeting of the American Electrophoresis Society (AES)

(365i) Nanofluidic Devices for DNA and Protein Pre-Concentration In High-Conductivity Media

Nathan Swami[1], Kuo-Tang Liao[1], Vasudha Chaurey[1] and Chiafu Chou[2], (1)Electrical & Computer Engineering Department, University of Virginia, Charlottesville, VA, (2)Academia Sinica, Taipei, Taiwan

Nanofluidic device design criterion for dielectrophoresis pre-concentration of functional ss-DNA and protein biomarker targets within media of high ionic strength, for eventual application towards clinical diagnostic systems of enhanced speed, sensitivity and selectivity are addressed in this presentation. The challenge of sensing rare numbers of biomarker target molecules against a background of high concentration of other matrix proteins can be addressed through selective biomarker pre-concentration. Dielectrophoresis (DEP) enables highly selective trapping of bio-particles based on the characteristic frequency response of the dielectric permittivity of the bio-particle versus that of the medium, and it has been extensively applied towards sorting of somewhat similar sized biological cells with differing dielectric frequency response. However, dielectrophoresis trapping of sub-50 nm bio-particles in high conductivity media is a challenge since temperature gradients due to localized Joule heating can lead to dissipative electrothermal flow. In this presentation, through device-level model simulations and experimental validation we aim to elucidate the criteria for designing devices for pre-concentration of sub-50 nm sized ss-DNA and ~5 nm scale proteins, so that DEP trapping is not hindered by dissipative electrothermal flow, while the sampled fluid volume is enhanced. DEP forces scale as the product of the field and its gradient (E.dE/dx or Del(E^2), while electrothermal forces scale as the second exponential power of electric field (σE^2), with no dependence on the electric field gradient. Hence, we seek to design constriction-based DEP devices that enhance electric field gradients over the net electric field intensity, to enhance DEP trapping forces over dissipative electrothemal forces. Through varying bio-particle size, media conductivity and constriction scaling factors, in positive and negative DEP modes, we present a scaling analysis for the balance of DEP versus electrothermal forces for nanofluidic designs that enable ss-DNA and protein pre-concentration. Finally, the design is applied towards the continuous monitoring of the enhancement of target binding kinetics as a function of degree of pre-concentration.

See more of this Session: DNA Analysis In Microfluidic and Nanofluidic Devices
See more of this Group/Topical: Topical 3: 2011 Annual Meeting of the American Electrophoresis Society (AES)

(366a) Characterization of Membrane Potential of Individual Mitochondria by Capillary Electrophoresis and Capillary Isoelectric Focusing

Gregory G. Wolken, Vratislav Kostal and Edgar A. Arriaga, Chemistry, University of Minnesota, Minneapolis, MN

Mitochondrial dysfunction has been associated with many diseases and the aging process. In cells and tissue, mitochondria are heterogeneous organelles forming subpopulations that exhibit differences in morphology, protein expression, respiratory capacity, and oxidative stress. These subpopulations of mitochondria may contribute differently to cellular dysfunction. Mitochondrial membrane potential is known to vary among mitochondria within a cell and among different subpopulations of damaged and dysfunctional mitochondria. Imaging techniques using mitochondrial-specific probes and fluorescence microscopy are useful to evaluate differences in mitochondrial membrane potential, but lack the ability to separate individual mitochondria for further characterization. We have used electrophoretic methods such as capillary electrophoresis with laser-induced fluorescence (CE-LIF) and capillary isoelectric focusing (cIEF) to separate individual mitochondria based on differences in their electrophoretic properties arising from variation in their surface composition. These techniques have the potential to simultaneously characterize mitochondrial electrophoretic properties and membrane potential when mitochondria are treated with membrane potential sensitive probes.

In this work, we are investigating the use of a ratiometric fluorescent probe sensitive for mitochondrial membrane potential, JC-1, for detection of individual mitochondria in CE and cIEF. Mitochondria from cultured L6 rat myoblast cells will be loaded with JC-1, injected into the capillary and separated with either CE or cIEF. Migration time and fluorescence from two channels will be recorded for individual mitochondria; this data will be used to determine electrophoretic mobility (CE) or isoelectric point (cIEF), and mitochondrial membrane potential. Distributions of both measurements will be used to determine the dependence between membrane potential and the electrophoretic parameters. Information on the distributions of membrane potential and the potential purification of mitochondria using electromigration techniques can lead to better understanding of mitochondrial turnover and dysfunction during disease and aging. This work is funded by the National Institutes of Health (R01-AG20866).

See more of this Session: Electrokinetic Behavior of Microparticles and Nanoparticles: Fundamentals and Applications
See more of this Group/Topical: Topical 3: 2011 Annual Meeting of the American Electrophoresis Society (AES)

(366b) Sedimentation of Charged Colloids In Strong Gravitational Fields

Aditya S. Khair, Chemical Engineering, Carnegie Mellon University, Pittsburgh, PA

The sedimentation of an electrically charged micron-sized particle is a fundamental problem in colloid science. The vast majority of previous theoretical work has focused on the "linear-response" limit, wherein the gravitational field is sufficiently weak (in an appropriate dimensionless sense) to only slightly perturb the equilibrium ionic screening cloud surrounding the colloidal particle. In this talk, we examine the motion of a charged spherical colloid under a "strong" gravitational field, which is relevant to ultra-centriguation and "streaming flows" in microfluidics, for instance. Here, importantly, the screening cloud can be highly distorted. We compute the sedimentation velocity of the particle over a range of field strengths and screening cloud thicknesses, via numerical solution of the linearized (i.e. small zeta potential) Poisson-Nernst-Planck equations. Time permitting, we will discuss the related, but significantly more challenging, problem of electrophoresis of colloids in strong electric fields.

See more of this Session: Electrokinetic Behavior of Microparticles and Nanoparticles: Fundamentals and Applications
See more of this Group/Topical: Topical 3: 2011 Annual Meeting of the American Electrophoresis Society (AES)

(366c) Size Exclusion Electrofocusing In Nanochannels

Cornelius F. Ivory, Voiland School of Chemical Engineering and Bioengineering, Washington State University, Pullman, WA

The standard method for protein sizing, SDS-PAGE, is cumbersome, slow, difficult to automate, and requires that protein bands be excised from the gel prior to further analysis. Capillary and microchip formats for performing SDS-PAGE have emerged as alternatives to the "slab gel" format, but have failed to displace traditional SDS-PAGE because of the relatively long runtimes required and the need for a pre-concentration "stacking" step to increase resolution.

In this paper, we consider replacing SDS-PAGE with an electrofocusing method that separates proteins on the basis of their molecular weights. Proteins and other nanoparticles can be simultaneously separated and concentrated according to their size in channels and/or pores whose dimensions are several times larger than the particles. In porous media such as monoliths, size-based electrofocusing can be conducted if there is a gradient in the properties of the pores while open nano-channels require a gradient in the channel properties to do this.

This paper will discuss how size-based electrofocusing can be carried out in several different media, the resolving power that can be expected from this technique, and how mathematical models can be used to describe these separations.

See more of this Session: Electrokinetic Behavior of Microparticles and Nanoparticles: Fundamentals and Applications
See more of this Group/Topical: Topical 3: 2011 Annual Meeting of the American Electrophoresis Society (AES)

(366d) Directed Assembly of Colloidal Dimers Under Electric Field

Fuduo Ma and **Ning Wu**, Chemical Engineering, Colorado School of Mines, Golden, CO

Anisotropic colloids are colloids that have been tailored for their presence of anisotropy in chemical compositions, surface functionalities, geometric sizes, or shapes. Therefore, they can recognize each other and assemble themselves via directional and specific interactions, potentially making large numbers of close-packed and non-close-packed crystalline structures that the isotropic spheres cannot form. In this talk, we will present our results in the directed assembly of the simplest form of anisotropic particles, colloidal dimers, under electric field. A thin film of dimer suspension is sandwiched between two conducting substrates, and an electric field perpendicular to the substrate is applied. We have studied the responses of both individual dimers and their collective motions under DC and AC electric field.

We have found rich behaviors in terms of both equilibrium phases and transient states. Under a DC field, individual dimers always orient perpendicularly to the electric field (parallel to the substrate) and they can aggregate into amorphous or semi-crystalline solids when the field strength is high enough. Increasing the frequency of the electric field, however, enables individual dimer particles to rotate and re-orient themselves in the direction parallel to the electric field (perpendicular to the substrate, i.e., "standing" dimer). Large areas of "standing" dimer crystals can be created within a window of intermediate frequencies. At even higher frequencies, the lateral dipole-dipole repulsions between dimers become dominant so that vertical chains of dimers span between top and bottom substrates. In addition, other complex patterns and phase behaviors will be presented. The experimental observations can be explained based on the electrokinetic theory of non-spherical colloids in bulk and electro-hydrodynamic flow near the substrate. The observed two- dimensional crystals of aligned dimers indicates a promising route to making photonic crystals using colloidal dimers as the building blocks.

See more of this Session: Electrokinetic Behavior of Microparticles and Nanoparticles: Fundamentals and Applications
See more of this Group/Topical: Topical 3: 2011 Annual Meeting of the American Electrophoresis Society (AES)

(366e) Colloid Aggregation and Concentration Through Double Layer Polarization

Stuart J. Williams, Birck Nanotechnology Center, Purdue University, West Lafayette, IN

Electrokinetic mechanisms like electrophoresis, dielectrophoresis, and electro-osmotic flow have revolutionized lab-on-a-chip systems. Their advantages are numerous and have been utilized to precisely control biological media and enhance particle separation. This presentation demonstrates a novel AC electrokinetic method that can rapidly and continuously accumulate and sort micro- and nanoparticles. The mechanisms responsible for this technique are founded upon the polarization and relaxation of the colloids' electric double layer. Previously the author has demonstrated this technique, termed rapid electrokinetic patterning (REP), through optically controlled methods. Here, an improved, portable REP device is demonstrated capable of massively parallel colloid concentration and sorting. Unlike dielectrophoresis which is based on interfacial polarization, REP utilizes the polarization of the colloids' electric double layer to trap and sort particles, enabling a new electrokinetic manipulation methodology. The ability to concentrate and sort particles suspended in a fluid sample directly impacts biological lab-on-a-chip and national security efforts.

See more of this Session: Electrokinetic Behavior of Microparticles and Nanoparticles: Fundamentals and Applications
See more of this Group/Topical: Topical 3: 2011 Annual Meeting of the American Electrophoresis Society (AES)

(366f) Ensemble Average TIRM: Theory and Application In Imaging Amperometry

Reza M. Rock, Paul J. Sides and Dennis C. Prieve, Chemical Engineering, Carnegie Mellon University, Pittsburgh, PA

Rapid testing of electrocatalysts and corrosion resistant alloys accelerates discovery of promising new materials. Imaging amperometry, the deployment of colloidal particles as probes of the local current density, allows simultaneous electrochemical characterization of the entire composition space represented in an alloy film electrode. Previous work has shown that variations in particle-electrode distance for single particles in changing electric fields can be measured using total internal reflection microscopy (TIRM), and used to determine the electrochemical current passing around the particle, independent of electrical measurements. [1,2]

Improving this method to enable simultaneous measurements across inhomogeneous samples involves imaging evanescent wave scattering from a two-dimensional random ensemble of particles levitated above the working electrode in electrochemical TIRM experiments. Imaging individual particles for this application is infeasible at the low magnification levels needed to image an entire macroscopic (~1 square cm) sample. Mapping of electrochemical activity across the surface can be achieved by dividing the electrode surface into a mosaic of square "patch" areas 100 μ m to a side, each containing 15-30 particles. The measured intensity for each patch is the sum of the scattering from all of the particles present in that patch. Due to the non-symmetric nature of the potential energy well of the particles in the vicinity of the wall, there may be significant discrepancies between data collected for individual particles and ensemble-average data of those same particles.

Theory has been developed for analysis of ensemble-average TIRM data using elements of the single-particle theory and considerations necessary for ensembles. Conversion of scattering intensity to current density requires knowing the scattering intensity at the most probable height of the particles at open circuit, but the mean intensity is the value measured in ensemble average experiments. In some cases these two values are indistinguishable but this is not typical for many experimental conditions. The results of this recent study demonstrate how to correct for this difference. The adjustment to ensembles has been verified with simulation and experiment. The new theory predicts that the variation between the currents calculated from ensemble and single-particle scattering data is a strong function of the weight of the particles in the ensemble; this conclusion is supported by experimental and simulation results presented.

Patch-level electrochemical TIRM experiments were conducted with polystyrene beads levitated over a tin doped indium oxide electrode in alkaline solution, at open circuit and during cyclic voltammetry (CV) experiments probing the oxygen evolution reaction. The measured raw TIRM scattering intensities of particle ensembles appropriately increase and decrease with corresponding values of applied potential. Furthermore, the new theory clearly converts experimentally measured TIRM intensities to local current densities approximately equal to the current density measured by traditional methods. The focus in this work is on determining the correct physics governing the particle movements in order to maximize the fidelity of the derived current densities. These same experiments were modeled using Brownian dynamics simulations to examine the interaction between the scattering from individual particles and that of the ensemble average. These investigations represent important foundational steps towards the use of the macroscopic imaging ammeter.

1. P J Sides, C L Wirth, D C Prieve "An Imaging Ammeter for Electrochemical Measurements," *Electrochemical and Solid State Letters* **13** F10-F12 **(2010).**

2. CL Wirth, PJ Sides, DC Prieve "The Imaging Ammeter," *J Coll Interf. Sci.* **357** 1-12 (**2011**).

See more of this Session: Electrokinetic Behavior of Microparticles and Nanoparticles: Fundamentals and Applications
See more of this Group/Topical: Topical 3: 2011 Annual Meeting of the American Electrophoresis Society (AES)

(366g) Mapping AC Electroosmotic Flow At the Dielectrophoresis Crossover Frequency of a Colloidal Probe

Jingyu Wang and H. Daniel Ou-Yang, Physics, Lehigh University, Bethlehem, PA

The AC electroosmosis (ACEO) flow field in the vicinity of conducting electrodes is mapped by the measurement of Stokes-Einstein forces on an optically trapped colloidal particle. Dielectrophoresis (DEP) of the probe particle is eliminated by measuring at the particle's DEP crossover frequency. At this frequency the DEP force on the particle vanishes. Isolation of ACEO from DEP of the probe utilizes the fact that the ACEO flow velocity depends strongly on the electric field, which is spatially dependent, whereas the DEP crossover frequency, is independent of the electric field. Elimination of the particle's DEP permits unambiguous mapping of the ACEO velocity field.

See more of this Session: Electrokinetic Behavior of Microparticles and Nanoparticles: Fundamentals and Applications
See more of this Group/Topical: Topical 3: 2011 Annual Meeting of the American Electrophoresis Society (AES)

(366h) Remote Powering and Steering of Self-Propelling Microcircuits by Modulated Electric Field

Rachita Sharma and Orlin D. Velev, Department of Chemical and Biomolecular Engineering, North Carolina State University, Raleigh, NC

We have demonstrated a new class of self-propelling particles based on miniature semiconductor diodes powered by an external uniform alternating electric field [1]. The millimeter-sized diodes floating in water rectify the voltage induced between their electrodes. The resulting particle-localized electroosmotic flux can propel them in the direction of either the cathode or the anode depending on their surface charge. These motile particles suggest rudimentary solutions to the problems facing self-propelling microdevices, including harvesting power from external sources for propulsion and potential for a range of additional functions. The next step in this direction is developing means of steering the moving microdevices. We will present and discuss a novel technique that allows simple and effective on-demand steering of the self-propelling diodes. We control remotely the direction of their locomotion by electronically modifying the duty cycle of the applied AC field. The diodes change their direction of motion when a DC component (wave asymmetry) is introduced into the AC signal. The DC component leads to electroosmotic redistribution of the counterions near the diode surface. The interaction of the dipole across the diode with the dipole across the redistributed counterions and the diode surface exerts torque on the diode, causing its rotation. Thus, the reversal of the direction of the local electroosmotic flux, caused by field asymmetry, leads to reversal of the direction of particle locomotion. When the duty cycle change is lower than the one needed to completely rotate the diodes, the diodes exhibit sideways motion. This new principle of combined AC+DC signal driven control of the direction of motion of self-propelling diodes can find applications in MEMs and micro-robotics.

[1] S. T. Chang, V. N. Paunov, D. N. Petsev, O. D. Velev, *Nat Mater*, ***6***, 235-240 (2007).

See more of this Session: Electrokinetic Behavior of Microparticles and Nanoparticles: Fundamentals and Applications

See more of this Group/Topical: Topical 3: 2011 Annual Meeting of the American Electrophoresis Society (AES)

(419a) Characterizing the Dielectric Properties of Human Mesenchymal Stem Cells Using a Quadrupole Microfluidic Device

Tayloria Adams, Chemical Engineering, Michigan Technological University, Houghton, MI, Brenda Ogle, Biomedical Engineering Department, University of Wisconsin, Madison, WI and Adrienne Minerick, Department of Chemical Engineering, Michigan Technological University, Houghton, MI

Characterizing the dielectrophoretic response of stem cells at various stages of growth and in the presence of promoters is an important first step in designing an electrokinetic microdevice to separate stem cell types [1]. Dielectrophoresis is a technique utilizing nonuniform electric field to polarize cells based on the polarizability and dielectric properties of their membrane, cytosol, and other structurally dominant organelles. The DEP force is also tunable via adjustments to the electric field magnitude and shape and the surrounding medium [2, 3]. Hemapoetic stem cell identification has been accomplished via antibody tagging of unique cell-surface antigens followed by flow cytometry and fluorescence-activated cell sorting, but adaptation to other types of stem cells has been limited [1]. Therefore, there is a need for an identification techique that does not require tagging and is versatile enough to identify and separate all types of stem cells. Dielectrophoretic separations are advantegous because identification of stem cells can be coupled with separation in a microfluidic device. In this work, a microfluidic device with gold quadrupole electrodes, microchannel and a pumping system will be used to characterize the DEP response of human mesenchymal stem cells. Mesenchymal stem cells are multipotent adult stem cells and can differentiate into osteoblasts, adipocytes, chondrocytes, astrocytes, and myoblasts based on environmental promoters [4]. The mesenchymal stem cells will be characterized at frequencies ranging from 100kHz to 80MHz in a medium with conductivity ranging between 0.01 and 0.5 S/m[t1] . These experiments will map out the cross-over frequency of the human mesenchymal stem cells in different stages of growth and in the presence of promoters. COMSOL simulations and MATLAB will be used to fit the data to a single or multishelled spherical DEP polarization model in order to back out structural polarizability and conductivities of components of the human mesenchymal stem cells. The broader implications of electrokinetically identifying differences in stem cells lie in purification and control for tissue engineering and cell therapy applications [4].

References:

1. Flanagan, L.A., Lu, J., Wang, L., Marchenko, S.A., Jeon, N.L., Lee, A.P., and Monuki, E., *Stem Cells*, **26**, 656-665, 2008.
2. Hawkins, B.G., Huang, C., Arasanipalai, S., and Kirby, B.J., *Analytical Chemistry*, **83**, 3507-3515, 2011.
3. Kirby, B.J., Micro- and Nanoscale Fluid Mechanics: Transport in Microfluidic Devices, New York: Cambridge University Press, 2010.
4. Wu, H.W., Lin, C.C., and Lee, G.B., *Biomicrofluidics*, **5**, 013401-013426 2011.

See more of this Session: Poster Session for the American Electrophoresis Society
See more of this Group/Topical: Topical 3: 2011 Annual Meeting of the American Electrophoresis Society (AES)

(419b) Monitoring Hydroxyl Radical Production In Mitochondria with Micellar Electrokinetic Chromatography

Margaret A. Donoghue and Edgar A. Arriaga, Chemistry, University of Minnesota, Minneapolis, MN

Reactive oxygen species (ROS) are thought to be a significant contributing factor to the aging process at a cellular level. Of particular interest are the ROS generated by the mitochondria, as research has pegged these to be a significant source of oxidative damage. Current methodology to investigate ROS production is heavily biased towards fluorescence microscopy. This method, however, lacks low limits of detection and the ability to distinguish between components fluorescing at the same wavelength. To overcome these drawbacks, we are developing sensitive capillary electrophoresis techniques with laser-induced fluorescence detection (CE-LIF) for ROS analysis. The treatment of cells and mitochondria with probes selective for a ROS of interest and subsequent separation will enable us to monitor low levels of specific ROS present in biological samples. The current focus of our work is the use of 2-[6-(4 - hydroxy)phenoxy-3*H*-xanthen-3-on-9-yl]benzoic acid (HPF) to monitor the production of hydroxyl radicals. In vitro controls have verified the selectivity of the probe, and the analysis of isolated mitochondria from L6 myoblasts shows our ability to measure basal and stimulated levels of hydroxyl radicals. The initial CE-LIF method development will be presented here, as well as results from control mitochondria. The application of HPF combined with CE-LIF to look at hydroxyl radical levels in mitochondria isolated from adipocytes will also be discussed.

See more of this Session: Poster Session for the American Electrophoresis Society
See more of this Group/Topical: Topical 3: 2011 Annual Meeting of the American Electrophoresis Society (AES)

(419c) Separation of Particles Using Contactless Dielectrophoresis and Electrokinetic Flow

Michael B. Sano[1], Roberto C. Gallo-Villanueva[2], Harsha Kittur[3] and Rafael Davalos[3], (1)School of Biomedical Engineering & Sciences, Virginia Polytechnic Institute and State University, Blacksburg, VA, (2)BioMEMS Reseach Chair, Tecnologico de Monterrey, Monterrey, Mexico, (3)School of Biomedical Engineering and Sciences, Virginia Polytechnic Institute and State University, Blacksburg, VA

Dielectrophoresis (DEP), the motion of a particle in a non-uniform electric field, has become a robust method for analyzing nano-particles, cells, viruses, and DNA based on their physical and electrical properties. A new technique, contactless Dielectrophoresis (cDEP), isolates cells from contact with the electrodes. This is achieved by using fluid electrodes which are isolated from the sample channel by thin insulating membranes. The absence of contact between electrodes and the sample fluid inside the channel prevents bubble formation and avoids any contaminating effects the electrodes may have on the sample. This technique has demonstrated the ability to isolate THP-1 human leukemia monocytes from a heterogeneous mixture of live and dead cells as well as discriminate between cells of different metastatic potential. Recently, this technique has been improved for operation at frequencies as low as 1 kHz, allowing us to manipulate cells using both positive and negative DEP. Typically, this technique relies on the use of pressure driven flow which creates a non-uniform flow velocity and requires the use of a syringe pump. This work presents the analysis and evaluation of the combination of cDEP with electrokinetic flow to isolate and enrich particles.

See more of this Session: Poster Session for the American Electrophoresis Society
See more of this Group/Topical: Topical 3: 2011 Annual Meeting of the American Electrophoresis Society (AES)

(419d) Mixing Enhancement in Microfluidic Devices Using Contactless Dielectrophoresis (cDEP)

Alireza Salmanzadeh, Biomedical Engineering and Sciences, Virginia Tech, Blacksburg, VA, Hadi Shafiee, Virginia Tech, Blacksburg, VA, Michael Sano, Engineering Science and Mechanics, Virginia Tech., Blacksburg, VA, Rafael Davalos, School of Biomedical Engineering and Sciences, Virginia Polytechnic Institute and State University, Blacksburg, VA and Mark A. Stremler, Engineering Science and Mechanics, Virginia Tech, Blacksburg, VA

Mixing in microfluidic devices has an important role in numerous biological and chemical applications. Rapid mixing is necessary in many lab-on-a-chip (LOC) and Micro Total Analysis Systems (μTAS) devices for chemical processes, including specific applications such as chemical reactions, and for biological processes, such as enzyme reactions, DNA or RNA analysis, and protein folding.

Despite the small length scale of microdevices, rapid, efficient mixing is difficult to achieve. Due to small length scale the Reynolds number, the ratio between inertia and viscous forces, is small. Thus mixing without any intentional stretching and folding of interfaces is dominated by molecular diffusion, which takes a long time relative to the typical operating timescale of these microdevices. To address this need, various techniques for enhancing mixing in microdevices have been proposed, including both active and passive techniques that are based on phenomena such as chaotic advection. There are many active mixing techniques, including those that use pressure, magnetohydrodynamic, acoustic, or thermal disturbances to enhance mixing. Electrokinetic mixing has proven to be an efficient method for actively mixing solutions or microparticles.

DEP, the motion of a particle in a suspending medium due to the presence of a non-uniform electric field, has shown great potential for particle separation, manipulation, and identification of micro- and nano-scale particles. However, studies in the literature using DEP for mixing enhancement are fairly limited.

In this study, we have investigated mixing in microdevices based on a new technique, termed contactless dielectrophoresis (cDEP). In cDEP, an electric field is created in a microchannel by electrodes that are inserted into two side channels filled with conductive solution. These side channels are separated from the main channel by thin insulating barriers that exhibit a capacitive behavior. An electric field is induced in the main channel by applying an AC field across the barriers. Not having direct contact between the sample and the metallic electrodes eliminates issues that plague conventional DEP such as bubble formation, electrode delamination, and sample contamination. Furthermore, the fabrication process is relatively simple because it is not necessary to pattern micro-electrodes in the main channels. This method thus is well suited to traditional mass fabrication techniques such as hot embossing and injection molding.

We consider four variations of a device that consists of two "mixing chambers" located on opposite sides of a 50 mm or 100 mm rectangular microchannel. The chambers are either rectangular or semi-circular, and they are either placed symmetrically on opposite sides of the channel or are staggered with an axial spacing of 300 mm. In all cases, the depth of the chambers is 50 mm to match the channel depth.

In our cDEP device, there are electrode channels which are separated from the mixing chamber by 20 mm insulating barriers. The electrode channels are filled with a conductive solution, and electric field gradients are generated in the microchannel using electrodes that are inserted into two electrode channels. The capacitive nature of the barrier between the chambers and the electrode channels generates a nonuniform electric field when an AC signal is applied. In some locations, the electrode channels are far from the mixing chamber to prevent their electric effect on the mixing chamber, thereby creating an electric field similar to that given by three isolated electrodes.

Before applying the electric field, the two inlet streams flow side-by-side through the devices with negligible diffusive mixing. After applying the electric field, the beads suspended in the bottom fluid are manipulated by the induced DEP force. The mixing chambers exhibit a resulting secondary flow. This secondary flow is observed to stretch and fold the fluid, which is an important characteristic of rapid mixing. The stretching and folding in each device occurs between the symmetrically placed mixing chambers. In devices with two separate mixing chambers, the first mixing chamber begins the stretching and folding, and then the second chamber amplifies this stretching and folding and enhances mixing.

We quantified the effectiveness of these devices by evaluating mixing index, which is equal to 1 minus the ratio of the standard deviation of the concentration after mixing to the standard deviation of the concentration before mixing. We ran mixing experiments for voltages between 0 and 300 V_{rms} (0, 50, 100, 150, 200, 250, 300 V_{rms}) at a constant frequency of 600 kHz and a flow rate of 5 μL/hr. We found that both devices with rectangular mixing chambers achieve a maximum mixing index of 80%. However, the maximum mixing index obtained in circular mixing chamber was 57%. This occurs because the rectangular shape of mixing chamber decreases the velocity of particles inside the mixing chamber, which give more time to DEP to influence the particles and create mixing. However, circular mixing chambers cannot have the same effect on particle's trajectories; thus, we do not observe efficient mixing.

We have also considered the dependence of mixing on the flow rate for a fixed voltage and frequency. The mixing efficiency decreases by increasing the flow rate, which is equivalent to decreasing the flow time scale. The results also showed that a higher mixing index can be obtained at higher voltages for a given flow rate.

In addition to mixing beads with a fluid, these microdevices can also be used for mixing two fluids when one of those fluids contains beads. To prove this claim, we suspended beads in a water-based blue dye solution and pumped the sample into one of the microdevice inlets. Deionized water was pumped through the other inlet. After applying an AC signal of 300 V_{rms} and 600 kHz, the dye solution got mixed with the deionized water due to the secondary flow generated by the DEP force on the beads.

See more of this Session: Poster Session for the American Electrophoresis Society
See more of this Group/Topical: Topical 3: 2011 Annual Meeting of the American Electrophoresis Society (AES)

(419e) Rapid Estimation of the Most Probable Number (MPN) of Viable Bacteria In Water Samples

Sachidevi Puttaswamy, Byung-Doo Lee, Ashley Jurgensmeyer and Shramik Sengupta, Biological Engineering, University of Missouri, Columbia, MO

Various regulatory agencies such as the USDA, FDA and EPA require that the number of viable bacteria present in certain samples of food or water be less than a certain specified number. For instance the US Pasteurized Milk Ordinance requires "Grade A" pasteurized milk to have a total viable bacterial count of ≤ 20,000 CFU/ml, and the EPA requires that the concentration of coliforms in primary recreational water bodies not exceed 500 CFU/100ml (5 CFU/ml)

The enumeration of VIABLE bacteria is typically done using a statistical technique known as the Most Probable Number (MPN) method. The method involves progressively diluting an aliquot from the sample of interest into microbial growth media (repeated in triplicate) and keeping a track of which dilutions show evidence of the presence of even a single VIABLE bacterium. Currently, the latter (whether a vial contains any viable bacteria, or not) is determined by culturing the vials for 1-2 days, and looking for the development of turbidity in the sample (either manually, or using a spectrometer). It is known that if there is even a single viable bacterium present, it will give rise to hundreds of millions, which will turn the solution turbid.

Since the concentrations of bacteria in the sample have to rise to millions of CFU/ml before they can be detected via turbidity measurements, it takes a long time (typically 1-2 days) before the MPN can be determined. This delay has many adverse economic (and even, health) consequences.

In contrast, our method promises to deliver MPN results in less than 1 day. The principle underlying our method of detection is the polarizability of viable bacterial cells. In the presence of an alternating electric field, there occurs a build-up of charge at the membrane, causing the cells to act like capacitors. As the bacteria multiply in number, there will be a corresponding increase in the charge stored in the interior of the suspension (its "bulk capacitance"), and this increase in bulk capacitance over time serves as a "signature" for the presence of viable bacteria.

We were able to quantify viable *E. coli* in Tryptic Soy Broth (TSB) in <6 hours as opposed to 1-2 days taken by the current methods and also have a cut off time of ~10 hours which significantly reduces the wait time for industries.

This method can thus have a significant impact on many modern water and food quality applications.

(419f) Theoretical and Experimental Approach On Simultaneous and Continuous Separation and Concentration of Particles On An iDEP Platform

Hector Moncada-Hernandez[1], Alejandro Abarca-Blanco[2], Jos A. Nieto-Franzoni[2] and Blanca H. Lapizco-Encinas[3], (1)BioMEMS Research Chair, Instituto Tecnolgico y de Estudios Superiores de Monterrey, Monterrey, Mexico, (2)Instituto Tecnolgico y de Estudios Superiores de Monterrey, Monterrey, Mexico, (3)Chemical Engineering, Tennessee Technological University, Cookeville, TN

Dielectrophoresis (DEP) is an electrokinetic technique successfully employed for the manipulation of particles in microfluidic decives. DEP is defined as the movement of dielectric particles due to an induced polarization in the presence of non-uniform electric fields. In this work, non-uniform electric fields are achieved employing insulating structures between a set of electrodes. This approach is known as insulator-based dielectrophoresis (iDEP). In the present work, performance of a simultaneous and continuous separator and concentrator of a mixture of particles is studied theoretically and experimentally. A two-particles mixture of polystyrene beads with different sizes are employed to evaluate the performance of the separator and concentrator on a DC-iDEP platform device. A computational model was developed in order to predict particles pathlines and estimate trapping regions where particles are concentrated. Theoretical and experimental results are compared in the present study.

See more of this Session: Poster Session for the American Electrophoresis Society
See more of this Group/Topical: Topical 3: 2011 Annual Meeting of the American Electrophoresis Society (AES)

(419g) Dielectrophoretic Response of Perfluorocarbon Oil-Core, Chitosan, Poly-L-Lysine, CaPO4-Shell Nanoparticles

Chungja Yang, Chemical Engineering, Michigan Technological University, Houghton, MI, Chun-Jen Wu, Materials Science and Engineering, University of Utah, Salt Lake, UT, Agnes Ostafin, Materials Science and Engineering, University of Utah, Salt Lake City, UT, Sean Kirkpatrick, Biomedical Engineering, Michigan Technological University, Houghton, MI and Adrienne Minerick, Department of Chemical Engineering, Michigan Technological University, Houghton, MI

Dielectrophoretic Response of Perfluorocarbon oil-Core, Chitosan, Poly-L-lysine, $CaPO_4$-Shell Nanoparticles

Chungja Yang[a], Chun-Jen Wu[b], Agnes E. Ostafin[b], Sean Kirkpatrick[c], Adrienne R. Minerick[a]

[a]Chemical Engineering Department, Michigan Technological University

[b]Materials Science and Engineering Department, University of Utah

[c]Biomedical Engineering Department, Michigan Technological University

Dielectrophoresis (DEP) is useful to assemble, discriminate, and manipulate particles based on a particle's polarizability and interactions with a surrounding medium under non-uniform electric fields. Core-shell nanoparticles are traditionally synthesized for drug delivery [1], nanoreactors [2,3], and diagnostic markers [4], and can be constructed with various liquid core and various solid shell materials. These engineered nanoparticles facilitate systematic investigation of DEP polarization mechanisms and are much simpler than biological cells. Perfluorocarbon oil in a phospholipid micellular structure were used for the core, along with three shell materials (chitosan, Poly-L-lysine, and $CaPO_4$) which were chosen for their dielectric properties and known synthesis procedures. We report the frequency-dependent, and medium conductivity-dependent responses of ~250nm core-shell nanoparticles. Particle-particle interactions were quantified by tracking pearl-chain assembly of the nanoparticles in two dimensions. Experiments were conducted within a 100nl chamber housing 100um wide Au quadrupole electrodes spaced 25um apart. Frequencies from 100kHz to 80MHz at $10V_{pp}$ were tested and the frequency-dependent motion was quantified via video microscopy image intensity profiles in two perpendicular planes. Experimental COF results were compared with spherical shell models of the real part of the Clausius-Mossotti factor as a function of frequency, conductivity of the shells, and medium conductivity. This work exhibited two dimensional nanoparticle patterning into pearl-chains and then bands, which were subsequently probed for coherent light scattering via spectral techniques.

[1] Schmidt, Chem. Mater, 2004, 16, 4942-4947

[2] Chen, Nanotechnology, 2010, 21, 455701

[3] Chen, Nanotechnology, 2010, 21, 215503

[4] Georganopoulou, PNAS, 2005, 102, 7, 2273-2276

See more of this Session: Poster Session for the American Electrophoresis Society
See more of this Group/Topical: Topical 3: 2011 Annual Meeting of the American Electrophoresis Society (AES)

(419h) Facile Isolation and Recovery of Biological Molecules with Chemically Triggered Degradable Polyacryamide Gel Electrophoresis

Jun Sung Kang, Chandrasekhar R. Rajagopalan, Santoshkumar L. Khatwani, Kristen M. Wagner and T. Andrew Taton, Chemistry, University of Minnesota, Minneapolis, MN

Polyacrylamide gel electrophoresis (PAGE) has been widely used in the analysis and separation of biological molecules, including proteins, DNA and RNA. For further analysis and manipulation of separated biological molecules, it is sometimes necessary to recover them from polyacrylamide gels. Most commonly, biomolecules are isolated from polyacryamide either by diffusion or electroelution. The poor diffusion of large biomolecules makes the methods slow and low-yielding. An alternative approach that has been proposed to reduce recovery time and increase yield is to incorporate degradable crosslinkers into the gel. Although degradable electrophoresis gels have been used to isolate biological molecules with high yield, they often require very limiting, harsh conditions, such as, acidic or basic buffer systems. To solve these limitations, we have developed novel crosslinkers consisting of α-azido ether functionality which can be degraded by bio-orthogonal chemical triggers. Here, we show that large biological molecules such as plasmid DNA can be separated by PAGE containing the new crosslinker, N_3-EG_2, and liberated with high yield via a chemical trigger. In addition, we find that the degradation conditions do not affect the activity of the biological molecules. We believe that N_3-EG_2 crosslinkers could be widely applied to biological molecule recovery and combined with other technologies like proteomics.

See more of this Session: Poster Session for the American Electrophoresis Society
See more of this Group/Topical: Topical 3: 2011 Annual Meeting of the American Electrophoresis Society (AES)

(419i) Morphological Effects On the Electrostatic Potential In a Divergent and Convergent Channel for Microfluidic Applications

Parvin Golbayani[1], Abbas Motamedilamouki[1], Kevin T. Seale[2], Robert Sanders[3] and Pedro E. Arce[1], (1)Chemical Engineering, Tennessee Technological University, Cookeville, TN, (2)Biomedical Engineering, Vanderbilt University, Nashville, TN, (3)Office of Research and Graduate Studies, Tennessee Technological University, Cookeville, TN

A number of researchers have proposed combining electroosmotic flow (EOF) and pressure-driven flow as a means of controlling the motion and separation of bioparticles in diverging- converging microchannels[1]. In order to get better understanding of role of electrostatic potential on fluid motion, analytical study of the electrostatic potential in divergent and convergent channels has been conducted in this study. Distribution of electrostatic potential is given by the solution of 2D Poisson-Boltzmann Equation (PBE) with both long channel and Debye –Huckel approximations. As a product of the investigation, one can assess the behavior of the electrostatic potential inside of a convergent-divergent section. Three key parameters have been identified to describe the electrostatic potential behavior: the angle ($\propto$) of the convergent (or divergent) section (related to the walls of the channels) that handles the "magnitude" of the deviation with respect to a straight channel; the ratio of the wall potentials, R, which handles the symmetrical/non-symmetrical aspects of the electrostatic potential, and the ratio of the width to the length (γ) that controls the "shape" of the channel section. Results of this study will be shown by using a series of portraits that capture the key behaviors of the electrostatic potential with respect to the three parameters described above.

1-V. Gnanaraj, V. Mohan, B. Vellaikannan, " Numerical Investigation of Electroosmotic Flow in Convergent Divergent Micronozzle", In proceeding of the COMSOL Conference, Bangalore(2009).

See more of this Session: Poster Session for the American Electrophoresis Society
See more of this Group/Topical: Topical 3: 2011 Annual Meeting of the American Electrophoresis Society (AES)

(419j) Experimental Validation of a DNA Pre-Conditioning Strategy Developed Using Brownian Dynamics Simulations

Cheng-Han Li, Department of Chemical Engineering, National Taiwan University, Taipei, Taiwan and Chih-Chen Hsieh, Chemical Engineering Department, National Taiwan University, Taipei, Taiwan

We have developed a new pre-conditioning strategy to improve DNA stretching by electrokinetic flow in a microfluidic device using Brownian dynamics simulations. The core of the strategy is to "pre-stretch" DNA in the direction orthogonal to its final stretching direction. This allows DNA to be pre-stretched while requires no additional strain to be provided. Although we have shown in simulations that this strategy can improve the degree of DNA extension considerably even at medium Deborah number, an experimental validation is yet to be performed. In this study, experiments on the newly developed pre-conditioning strategy and some of its variation are performed. We compare in details the predictions and experimental results in the averaged maximum DNA extension, its distribution and the mechanism causing the preconditioning effect. It is found that the new strategy provides moderate improvement on the degree of DNA extension. The cause of the discrepancy between simulations and experiments will be discussed.

See more of this Session: Poster Session for the American Electrophoresis Society
See more of this Group/Topical: Topical 3: 2011 Annual Meeting of the American Electrophoresis Society (AES)

(419k) Analysis of Channel Morphology for Electrophoresis of Biomolecules: Effect of Axial and Orthogonal Fields

Jyothirmai J. Simhadri[1], Mario Oyanader[1], Holly A. Stretz[2] and Pedro E. Arce[1], (1)Chemical Engineering, Tennessee Technological University, Cookeville, TN, (2)Department of Chemical Engineering, Tennessee Technological University, Cookeville, TN

Research involving polymer gels with embedded nanoparticles of varying properties is quite attractive because of the multitude of potential applications, including separation of biomacromolecules for clinical diagnostics, tissue scaffold growth etc. Moreover, analysis of the effects of the pore length and diameters on the transport of biomolecules is an important aspect to be studied either analytically or computationally as shown by previous efforts [1]. In this research, we propose to computationally analyze different pore models (associated with gel materials) and study the effect of geometry on the transport properties. Results will illustrate, for example, the role of pore-scale in conjunction with the effect of secondary electrical fields.

References:

1. Trinh, S., B.R. Locke and P.E. Arce, *Separation and Purification Technology*, 15, 255 (1999).

See more of this Session: Poster Session for the American Electrophoresis Society
See more of this Group/Topical: Topical 3: 2011 Annual Meeting of the American Electrophoresis Society (AES)

DESIGN AND SIMULATION OF AN AUTOMATED RARE BLOOD CELL DETECTOR

Zhixi Qian[a], Paul W. Todd[b], and Thomas R. Hanley[a]

[a]Department of Chemical Engineering, Auburn University, AL 36849

[b]Techshot, Inc. 7200 Highway 150, Greenville, IN 47124

In recent years the demand for using circulating tumor cell (CTC) counts in cancer management has increased. With the ability to detect small numbers of rare cells in blood and the motivation to non-invasively monitor therapeutic progress, CTC detection technology would play an important diagnostic role. However, the current detection process can be performed only in specialized laboratories with the analysis requiring two to four skilled technicians. The device's price can reach seven figures, limiting the availability for cancer detection. Analysis requires weeks (or even months), making it difficult to detect disease at an early stage.

Our goal is to design a portable point-of-care device consisting of three syringes, a static mixer, a magnetic filter, a microfluidic "chip" and a waste receptor. Compared with current detection systems, our device is much smaller, lower in price, and requires only one operator. The device would be available for use in clinics, nursing homes or residences for detection of most cancers (pancreas, ovarian and melanoma) in an early stage.

Our recent research focuses on the simulation of the cell detector. CFD software is utilized to simulate the static mixer's operation, mixing the blood sample, buffer and reagents. CFD simulation results are then compared to the results of laboratory experiments. Following the static mixer simulation, the magnetic filter is simulated, incorporating magnetic particle motion. The results of the simulation are compared to experimental results with the results used to optimize the size and configuration of the magnet filter.

Key words: circulating tumor cells, computational fluid dynamics, magnetic separation, point-of-care treatment

(419m) Optimal Design of Microfluidic Capillary Networks for Rapid Gel Free DNA Separation

Max A. Fahrenkopf, B. Erik Ydstie and James W. Schneider, Chemical Engineering, Carnegie Mellon University, Pittsburgh, PA

10 to 100 times reductions in run time over the current state-of-the-art gel electrophoresis DNA length based separation are made possible through the mathematical optimization of a novel separation technique known as micelle end-labeled free solution electrophoresis (ELFSE). In gel electrophoresis the gel sieving matrix, necessary to make DNA velocity length dependent through friction, performs at typical lab speeds of 400 bases in 1 hour. The absence of a sieving matrix in aqueous solution will ensure that DNA of all lengths will migrate at the same velocity without another friction causing agent. In micelle ELFSE this friction is generated by a dilute low viscosity solution of micelles allowing for unique, superior trade-offs in operating conditions, like electric field strength, capillary length, and micelle size, necessary for large increases in throughput. These trade-offs depend on the separation platform. Capillary arrays and microfluidic devices are examined as candidates for the fastest separation mode.

Length based separation is achieved when DNA populations of different lengths migrate at different velocities. Small deviations in DNA population velocity cause ideal Dirac-delta signals to widen into Gaussian signals which can overlap and become unresolved. These small deviations in DNA velocity are due to micelle polydispersity, diffusion, and other mass transfer effects as described by a moment analysis which renders a model similar to the Van Deemter equation in chromatography to describe the Gaussian width. Micelle polydispersity is typically the largest signal widening effect which is reduced over time by time averaging through transient binding of the micelle to the end of alkylated DNA. Because of the necessity of DNA to sample different micelles over time, fast run times and high resolution are conflicting objectives that are difficult to mitigate without a rigorous optimization approach.

Non-linear programming techniques are utilized to solve the dual objective optimization problem. The model is non-convex and requires a global optimization code like BARON to generate the pareto front. BARON identifies standard form non-convexities to relax into a convex underestimate proceeding through a spatial branch and bound to reduce the gap between the convex underestimate with the non-convex local solution to finally converge to the global optimum. The model is written to expose these standard form non-convexities (bilinear, linear fractional, and concave univate terms) for fast convergence of BARON.

Typical lab scale separations are performed using either a capillary array in a bench top system or in a curved channel in a microfluidic device. Single capillary, parallel capillaries, serpentine microfluidics, and spiral microfluidics are examined; as well as finish-line versus full channel snap shot detection modes. Snap shot is the fastest detection mode possible as every population of DNA smaller than the length of read is guaranteed to be resolved in channel once the length of read is resolved, owning to the monotonic nature of the resolution. Electric field strength and channel length are found to be strong functions of the micelle polydispersity. A reduced parameter combination of the electric field, channel length, and micelle polydispersity is used to collapse the optimization results for different levels of polydispersity onto a single curve. Minimized run times are shown to have significant improvement over gel based separations. Specifically 27%, 74%, 84%, and 90% reduction in run time out to 400 bases over gel for single capillary, 2 parallel capillaries, serpentine microfluidics, and spiral microfludics respectively. Even more improvement may be realized through the use of an electro-osmotic counter flow to slow down fast moving, hardly resolved long DNA.

References

S. T. Grosser, J.M. Savard, Schneider, J.W, *Anal. Chem.* 2007, 79, 9513–9519

Ren, H., Karger, A. E., Oaks, F. Menchen, S., Slater, G. W., Drouin, G. *Electrophoresis*, 1999, 20, 2501-2509

Aris, R. *Proc. R. Soc. London Ser. A*, 1956, 235, 67-77

Sahinidis, N. V., Tawarmalani, M, *Mathematical Programming*, 2005, 103, 2, 225–249

Smith, E.M.B, Pantelides, C.C., *Comp. and Chem. Eng.*, 1999, 23, 457-478

Quesada, I., Grossmann, I. E., *J. Global Optim.*, 1995, 6, 39-76

L. C. McCormick, G. W. Slater, *Electrophoresis*, 2006, 27, 1693-1701

See more of this Session: Poster Session for the American Electrophoresis Society
See more of this Group/Topical: Topical 3: 2011 Annual Meeting of the American Electrophoresis Society (AES)

(419n) Degradation of Kappa-Casein In Aqueous Solutions by Pulsed Corona- Based High Oxidation Method: Further Studies

Chinyere P. Mbachu[1], Maria F. Ibaez[1], Tyler J. Cosby[1], Dennis B. George[2] and Pedro E. Arce[1], (1) Chemical Engineering, Tennessee Technological University, Cookeville, TN, (2)Center for the Management, Utilization, and Protection of Water Resources, Tennessee Technological University, Cookeville, TN

One of the key physical parameters for successful degradation of humic acid, for the case of direct Pulsed Corona Discharges was determined to be the electrode-to-electrode gap. In this, we are texting this parameter in the degradation of kappa-casein and reporting results of its effects on degradation. Non-thermal high oxidation method, a novel oxidation technique based on localized electrical discharge has been used successfully to degrade high molecular weight organic compounds (dye solutions, humic acids and fulvic acids) before. In this research, a pulsed corona discharge was applied to reduce kappa-casein in aqueous solutions and the effects of pH, and frequency were investigated.

See more of this Session: Poster Session for the American Electrophoresis Society
See more of this Group/Topical: Topical 3: 2011 Annual Meeting of the American Electrophoresis Society (AES)

(424a) The Use of Microchannel Electrophoresis to Understand Amyloid Aggregation

Elizabeth Pryor, Chemical Engineering, University of Arkansas, Fayetteville, AR, Melissa A. Moss, Chemical Engineering, University of South Carolina, Columbia, SC and Christa N. Hestekin, Ralph E. Martin Department of Chemical Engineering, University of Arkansas, Fayetteville, AR

Amyloids are involved in a variety of diseases including Alzheimer's Disease, Parkinson's disease, and diabetes. A number of technical challenges exist for the quantitative analysis of the different sizes of amyloid proteins which are present during the early stages of aggregation, therefore requiring new technologies to be developed for enhanced protein separations. Microchannel electrophoretic techniques have emerged as powerful tools for the quantitative analysis of proteins. In this study, we investigated the ability of CE to monitor the early stages of amyloid aggregation using insulin and amyloid beta as model amyloid-forming proteins. The effect of solutions conditions, including pH and ionic strength, on lag time and aggregate size have been determined.

See more of this Session: Advances In Electrophoretic Protein Separation and Analysis
See more of this Group/Topical: Topical 3: 2011 Annual Meeting of the American Electrophoresis Society (AES)

(424b) Insulator-Based Dielectrophoresis Applied to Immunoglobuline G and Bovine Serum Albumin Concentration

Alexandra Ros, Fernanda Camacho-Alanis, Asuka Nakano and Tzu-Chiao Chao, Department of Chemistry and Biochemistry, Arizona State University, Tempe, AZ

Dielectrophoresis (DEP) is a well established technique to manipulate particles and biological objects such as bacteria and mammalian cells. Its capability of sorting, fractionation and separation make it also very attractive as a new method for biomolecule separation without gel- matrices or chromatographic phases. Here, we exploit dielectrophoresis of the proteins immunoglobulin G (IgG) and bovine serum albumin (BSA) in a tailored microfluidic device and demonstrate concentration of proteins due to DEP streaming behavior.

An insulator-based microfluidic device serves as the basis of our DEP studies. Insulating posts fabricated with standard photolithographic and softlithographic methods create electric field gradients upon application of external fields. Those electric field gradients are used to manipulate proteins by DEP. First, we investigated the electric field distribution in the obtained microstructures. From those, the migration behavior of proteins was simulated with a numerical model that combined DEP forces with electrokinetic and diffusional transport mechanisms. We could demonstrate, that proteins exhibiting positive DEP behavior are concentrated in streamlines. In contrary to this, proteins with negative DEP show concentration depletion in the same streaming regions.

Next, we found experimentally that both proteins – IgG and BSA – showed DEP streaming behavior. This experimentally observed streaming behavior was in excellent qualitative agreement with numerical simulations in the case of positive DEP. A concentration factor of up to 24% was obtained for IgG, indicating the potential of protein concentration via DEP in microfluidic devices. Furthermore, our experiments demonstrate that protein aggregation in iDEP can be effectively reduced by the addition of zwitterionic detergents.

In the future, we will extend protein iDEP manipulation to combined micro- and nanostructured devices, in which we expect increased concentration behavior due to increased electric field gradients. Our goal is further to exploit protein DEP and the underlying polarization behavior quantitatively to apply DEP of proteins for separation and fractionation.

Dielectrophoresis (DEP) is a well established technique to manipulate particles and biological objects such as bacteria and mammalian cells. Its capability of sorting, fractionation and separation make it also very attractive as a new method for biomolecule separation without gel- matrices or chromatographic phases. Here, we exploit dielectrophoresis of the proteins immunoglobulin G (IgG) and bovine serum albumin (BSA) in a tailored microfluidic device and demonstrate concentration of proteins due to DEP streaming behavior.

An insulator-based microfluidic device serves as the basis of our DEP studies. Insulating posts fabricated with standard photolithographic and softlithographic methods create electric field gradients upon application of external fields. Those electric field gradients are used to manipulate proteins by DEP. First, we investigated the electric field distribution in the obtained microstructures. From those, the migration behavior of proteins was simulated with a numerical model that combined DEP forces with electrokinetic and diffusional transport mechanisms. We could demonstrate, that proteins exhibiting positive DEP behavior are concentrated in streamlines. In contrary to this, proteins with negative DEP show concentration depletion in the same streaming regions.

Next, we found experimentally that both proteins – IgG and BSA – showed DEP streaming behavior. This experimentally observed streaming behavior was in excellent qualitative agreement with numerical simulations in the case of positive DEP. A concentration factor of up to 24% was obtained for IgG, indicating the potential of protein concentration via DEP in microfluidic devices. Furthermore, our experiments demonstrate that protein aggregation in iDEP can be effectively reduced by the addition of zwitterionic detergents.

In the future, we will extend protein iDEP manipulation to combined micro- and nanostructured devices, in which we expect increased concentration behavior due to increased electric field gradients. Our goal is further to exploit protein DEP and the underlying polarization behavior quantitatively to apply DEP of proteins for separation and fractionation.

See more of this Session: Advances In Electrophoretic Protein Separation and Analysis
See more of this Group/Topical: Topical 3: 2011 Annual Meeting of the American Electrophoresis Society (AES)

(424c) PILOT_PROTEIN: A High-Throughput Method for In Silico Discovery of Peptides, Proteins, and Post-Translational Modifications

Richard Baliban[1], Peter A. DiMaggio Jr.[2], Zukui Li[1], Mariana Plazas-Mayorca[3], Benjamin A. Garcia[2] and Christodoulos A. Floudas[1], (1)Chemical and Biological Engineering, Princeton University, Princeton, NJ, (2)Molecular Biology, Princeton University, Princeton, NJ, (3)Department of Chemistry, Princeton Universirty, Princeton, NJ

A fundamental problem in proteomics is that of *modified protein identification*, which corresponds to determining sequence of a protein along with all amino acid modifications that were added to the protein after it was constructed *in vivo* (post-translational modifications or PTMs). We present a comprehensive set of tools that can address this problem using mixed-integer linear optimization (MILP) and tandem mass spectrometry. These tools have been integrated into a singular webtool that is freely available to the scientific community.

MS/MS spectra are initially analyzed using the de novo sequencing algorithm PILOT [1]. PILOT can rigorously guarantee a rank-ordered list of optimal candidate sequences without complete enumeration of all possible sequences. To utilize the strengths of a database routine, a hybrid de novo/database routine using local sequence alignment, PILOT_SEQUEL [2], was developed to input the results of the PILOT method and query these sequences against a database to find peptide matches.

A comprehensive unmodified protein list is constructed using the novel method PILOT_PROTEIN [3]. PILOT_PROTEIN will use the scored peptides of PILOT_SEQUEL to generate the protein list and incorporates a peptide clustering routine to reduce false positives. Using a known list of proteins, either an untargeted or a targeted PTM search can be performed to determine the set of PTM types and sites that best explains the experimental data [4]. If the proteins are known to be highly modified, a MILP approach may be used to determine the identification and quantification of all proteins that may exist within each MS/MS. The resulting output from the above routines is a complete modified protein list.

To verify the protein prediction capability of each method, several comparative studies were performed against state-of-the-art algorithms using data sets from a variety of MS/MS instruments and fragmentation types. Specifically, the peptide prediction accuracy of PILOT and PILOT_SEQUEL, the protein identification accuracy of PILOT_PROTEIN, and the PTM prediction accuracy of PILOT_PTM were analyzed using over 170 LC-MS/MS data sets from the Standard Protein Mix Database [5] comprising a total of 1.5 million MS/MS. Each of the methods produces superior predictive capability when compared to other algorithms and maintains these results across all of the test data sets.

[1] P. A. DiMaggio Jr. and C. A. Floudas. De novo peptide identification via tandem mass spectrometry

and integer linear. Anal. Chem., 79:1433-1446, 2007.

[2] P. A. DiMaggio Jr., B. Lu, J. R. Yates III, and C. A. Floudas. A Hybrid Method for Peptide Identification Using Integer Linear Optimization, Local Database Search, and Quadrupole Time-of-Flight or OrbiTrap Tandem Mass Spectrometry. J. Proteome Res., 7(4):1584–1593, 2008.

[3] R. C. Baliban, P. A. DiMaggio Jr., ZuKui Li., M. D. Plazas-Mayorca, B. A Garcia, and C. A. Floudas. Identification of modified and unmodified proteins via high-resolution tandem mass spectrometry and mixed-integer linear optimization. Mol. Cell Proteomics, submitted.

[4] R. C. Baliban, P. A. DiMaggio Jr., M. D. Plazas-Mayorca, N. L. Young, B. A Garcia, and C. A. Floudas. A Novel Method for Untargeted Post-Translational Modification Identification Using Integer Linear Optimization and Tandem Mass Spectrometry. Mol. Cell Proteomics, 9:764-779, 2010.

[5] J. Klimer, J. S. Eddes, L. Hohmann, J. Jackson, A. Peterson, S. Letarte, P R. Gafken, J. E. Katz, P. Mallick, H. Lee, A. Schmidt, R. Ossola, J. K. Eng, R. Aebersold, and D. B. Martin. The Standard Protein Mix Database: A Diverse Data Set To Assist in the Production of Improved Peptide and Protein Identification Software Tools. J. Proteome Res., 7(1):96–103, 2008.

See more of this Session: Advances In Electrophoretic Protein Separation and Analysis
See more of this Group/Topical: Topical 3: 2011 Annual Meeting of the American Electrophoresis Society (AES)

(424d) A Multi-Dimensional Approach for Comprehensive LC-MS/MS Identification and Quantitation of Highly-Modified Protein Systems

Peter A. DiMaggio Jr., Nicolas L. Young and Benjamin A. Garcia, Department of Molecular Biology, Princeton University, Princeton, NJ

Existing computational approaches for identifying and quantitating proteins with multiple post-translational modifications (PTMs) from LC-MS/MS data generally decouple this complex but complementary data structure into independent sub-problems for tractability purposes. However, this approach ignores important chromatographic relationships that exist in the *m/z* and time dimension and can only quantitate a modified form if it has been "positively identified" using tandem MS. This is a severe limitation, as tandem mass spectra (of varying quality, which is dependent on many factors) are only available for a fraction of the peptides that are actually present in a biological sample due the limited dynamic range of the mass spectrometer. Therefore, existing strategies fail to provide robust and comprehensive biological readouts as they cannot detect lower abundance forms nor resolve co-eluting isobaric peptides for which only partial, mixed or no tandem MS information is available.

We have developed a method for targeted protein systems that solves a much larger-scale problem to simultaneously annotate and quantify all of the peptides present in a LC-MS/MS run. The approach can utilize both label-free and isotopically-labeled data from a variety of sources, including metabolically-labeled amino acids and/or their transiently-labeled post-translational modifications. The key concept of our novel method is that chromatographic information (such as peak shape, isotopic distribution and relative elution time with respect to physically-related peptides) is directly incorporated into the MS2 identification and MS1 quantitation problems. By formulating this as an optimization problem, all modes of this information are simultaneously considered and the resulting model, which is more representative of the actual large-scale LC-MS/MS data structure, is solved to global optimality to identify and quantify ALL of the targeted peptides at once. Furthermore, the method is able to deconvolve co-eluting isobaric peptides present in mixed tandem MS (i.e., spectra containing the fragment ions of more than one modified form) and make accurate and robust identifications to incomplete tandem MS. To motivate the main ideas of our approach, specific examples will be provided to illustrate how unambiguous annotation and quantitation can only be accomplished for a particular peptide when additional information outside of the tandem MS is considered.

The utility of our method will be demonstrated and compared to existing quantitation algorithms using both bottom up and middle down LC-MS/MS data corresponding to mixtures of hyper-modified proteins from several biological samples. The systems presented in this work were selected to highlight the understudied yet ubiquitous existence of multiply-modified proteins and the functional implications of their PTMs in the molecular processes of eukaryotes. These hyper-modified protein systems include: (1)

chromatin-related proteins, such as histones, in which several hundred modified isoforms are present per sample and correlate directly with specific nuclear events; (2) proteins essential for regulating telemore replication, in which differential phosphorylation simultaneously occurs on several residues; and (3) enzymatic proteins, where the modification state of the functional protein provides important insight into its structure. These systems further highlight the critical importance of obtaining quantitative measurements for all modified isoforms of a peptide, as lower abundance forms for which no tandem MS are available are often the most biologically relevant but are not detected by other methods. Lastly, it will be illustrated how the robust and comprehensive readouts generated by our algorithm can be used to reveal biological insight into these processes.

See more of this Session: Advances In Electrophoretic Protein Separation and Analysis
See more of this Group/Topical: Topical 3: 2011 Annual Meeting of the American Electrophoresis Society (AES)

(424e) Fractionation of Complex Peptide Mixtures by IEF: Comparison of off-Gel with In-Gel Separations

Tom Berkelman[1], Sricharan Bandhakavi[1], Matthew Stone[2], Annett Hahn-Windgassen[1] and Aran Paulus[1], (1)Bio-Rad Laboratories, Hercules, CA, (2)University of Minnesota

Shotgun proteomic analysis involves the separation of proteolytically digested samples by reverse-phase liquid chromatography followed by mass spectrometry with electrospray ionization (LC-MS). This technique can detect thousands of individual peptides while identifying their protein of origin. It thus provides a powerful profiling technique for proteins present in a biological sample. Coupling this technique with an upstream, orthogonal separation technique yielding several fractions to be analyzed separately results in higher total numbers of identified peptides greater depth of proteome coverage. Isoelectric focusing (IEF) in an immobilized pH gradient (IPG) is a common upstream fractionation method used for this purpose. The IEF separation may be performed "Off-gel" in which the digest is electrophoretically fractionated through a series of sample cups placed in contact with an IPG gel (1), or "In-gel" in which the digest is incorporated into an IPG gel by rehydration and separated within the gel matrix (2). Off-gel separated peptides are recovered directly from the sample cups, while in-gel separated peptides are eluted from excised sections of IPG gel. This study was undertaken in order to compare the two methods in terms of overall speed and simplicity of the workflow, depth of proteome coverage provided, and the resolution of the separation as judged by the percentage of peptides restricted to a single fraction.

The resolution of the two different separation techniques was similar; however the in-gel separation resulted in approximately 20% more peptides detected. The in-gel separation was more rapid, but required additional time for peptide elution following the separation. An additional advantage of the in-gel separation is greater flexibility in determining the pH range covered by each individual fraction. In-gel peptide IEF fractionation can be performed with relatively inexpensive equipment and offers distinct benefits compared to off-gel peptide IEF fractionation.

(1) Hörth, P. et al. (2006). Molecular and Cellular Proteomics **5**:1968-1974

(2) Cargile, B.J et al. (2005). Journal of Biomolecular Techniques **16**:181-189

See more of this Session: Advances In Electrophoretic Protein Separation and Analysis
See more of this Group/Topical: Topical 3: 2011 Annual Meeting of the American Electrophoresis Society (AES)

(424f) Identification of Tyrosine Kinases and Their Substrates In Human Tumor Samples by 2D Gel Phosphotyrosine Western Blotting

Nancy Kendrick[1], Matt Hoelter[1], Jon Johansen[1], Christina Rose[1] and Mary Ann Gawinowicz[2], (1) Kendrick Labs Inc, Madison, WI, (2)Protein Core Facility, Columbia University, New York, NY

Tyrosine kinases (TK) are known to initiate mitotic cell division and often drive cancer growth. Tyrosine kinase inhibitors (TKI), Gleevec for example, are a class of cancer drugs that sometimes have spectacular success at inhibiting cancer, but other times fail completely. High throughput genomic methods have been unsuccessful at predicting which of the many available TKIs match to which cancers. One problem with genomic methods is that protein levels are often not proportional to mRNA levels. Pre (microRNA) and post-translation mechanisms exist at many levels in mammalian cells to regulate protein expression. Here we describe progress in direct identification of phospho-tyrosinylated proteins in human tumor tissue using the method of 2D gel phosphotyrosine Western blotting followed by mass spectrometry. Since tyrosine kinases often self-phosphorylate, the kinases themselves also light up. Flash-frozen lung, liver, kidney and pancreatic tumor samples along with normal tissue controls were purchased from a human tissue bank. The samples were homogenized and the proteins resolved by carrier ampholine 2D gel electrophoresis followed by western blotting with a PY20 anti-phosphotyrosine antibody. The ECL films were matched to duplicate Coomassie gels and spots cut from the latter for protein identification by mass spectrometry. Differences and similarities between the tumor sample western blot patterns will be discussed.

See more of this Session: Advances In Electrophoretic Protein Separation and Analysis
See more of this Group/Topical: Topical 3: 2011 Annual Meeting of the American Electrophoresis Society (AES)

(424g) The Charge of Green Fluorescent Protein, and Its Effect On Advanced Electrophoretic Separations

Frank Jahnke and Penny Ross, Sonata Biosciences, Inc., Auburn, CA

The charge of proteins as they vary with pH determines in large part how one protein may be separated from another. The isoelectric point of many proteins is known, and this information can be used to guide development of an advanced electrophoretic separations method. The isoelectric point alone, however, gives no information on how the protein charge changes. Often simple group contribution methods that sum the pKa of the amino acid residues are used to estimate the charge of a protein as pH is varied. It is recognized that the group contribution approach has many limitations, though it is rare that they are explored in detail for any particular protein.

In this presentation the charge of Green Fluorescent Protein, a widely-used fluorescent reporter, is presented as a function of pH over the range of 3 to 10. The charge measured by experiment differs drastically from what would be expected from its amino acid residues. Probable reasons for the difference, and its effect on electrophoretic separations, is discussed.

See more of this Session: Advances In Electrophoretic Protein Separation and Analysis
See more of this Group/Topical: Topical 3: 2011 Annual Meeting of the American Electrophoresis Society (AES)

Isolating Cancer Cells from Blood Using Contactless Dielectrophoresis (cDEP)

Alireza Salmanzadeh,[1,2] Mike B. Sano,[1] Hadi Shafiee,[3] Mark A. Stremler,[2] and Rafael V. Davalos[1,2,]*

Email:davalos@vt.edu

[1]School of Biomedical Engineering and Sciences, Virginia Tech-Wake Forest University, Blacksburg, VA

[2]Department of Engineering Science and Mechanics, Virginia Tech, Blacksburg, VA

[3]Biomedical Engineering Center, Department of Medicine, Harvard Medical School, Boston, MA

Abstract

Cell enrichment is an important process in many medical and biological applications including early cancer detection and individualized medicine. Circulating tumor cells (CTCs) are rare cell population which are shed from tumors and disseminated into the vasculature. Isolation and characterization of these cells is a challenging step in the development of early cancer detection techniques[1]. Dielectrophoresis (DEP) is a phenomenon in which cells are attracted to regions of high non-uniformity in an electric field. This force is dependent on each cell types unique biophysical and electrical properties[3]. We have developed a technique, contactless dielectrophoresis (cDEP)[2], which exploits these forces to isolate rare cell populations. In this study, we successfully isolated leukemia, prostate, and breast cancer cells from blood. The devices examined achieved an order of magnitude increase in throughput over previously reported cDEP devices, improving total throughput from 0.02 mL/hr[4] to 0.2 mL/hr. At this elevated throughput, the ability to isolate 1 cancer cell per 10^7 hematologic cells with an efficiency of 96% is demonstrated.

Introduction

There are numerous scenarios in cancer biology in which it is highly desirable to isolate subpopulations of cells. These applications include individualized medicine, metastatic analysis and detection of circulating tumor cells (CTCs) from blood. In traditional cell isolation technologies, such as flow cytometry[11], chemically functionalized pillar-based microchips[12], or magnetic bead cell separation[13], there is an inherent trade-off between throughput and purity limits.

Dielectrophoresis (DEP) is a phenomenon in which cells are attracted to regions of high electric field non-uniformity within an suspending medium. This force is dependent on each cell types unique biophysical and electrical properties[3]. This force has been successfully utilized on the microscale to separate of neuroblastoma cells from HTB glioma cells[5], cervical carcinoma cells[6], K562 human CML cells[7], live yeast cells from dead[8], and mammalian cells based on their cell-cycle phase[9, 10]. Typically, dielectrophoretic devices consist of metal electrodes patterned on the bottom of a microfluidic channel. The geometry of the electrodes is controlled to produce regions of non-uniformity in the electric field. Alternatively, insulator based dielectrophoresis (iDEP) uses insulating structures within a microfluidic channel to create non-uniformities in the electric field when metal electrodes are inserted into the inlet and outlet of the channel.

Contactless dielectrophoresis (cDEP) is a new adaption of DEP technologies. In this technique, fluid electrodes are isolated from a sample channel by a thin insulating membrane. When an AC voltage is applied across the fluid electrodes they become capacitivly coupled with the sample channel. This action creates a spatially non-uniform electric field within the sample channel providing sufficient conditions for cell manipulation. This technique eliminates direct contact between the sample fluid and the electrodes while maintaining the simplicity of fabrication associated with iDEP devices.

It has been suggested that for patients with isolated cancerous tumors, there are fewer than ten circulating cancer cells per milliliter of whole blood. This is an exceedingly small quantity in contrast to the number of hematologic cells, which in the same sample volume, can be 10^9 cells or greater. The nature of this challenge requires that devices have high volumetric throughput, removal efficiency, and selectivity. In this study, we present a cDEP device consisting of two wide sample channels in parallel which achieved an order of magnitude increase in throughput over previously reported cDEP devices, improving total throughput from 0.02 mL/hr to 0.2 mL/hr. At this elevated throughput, the ability to isolate 1 cancer cell per 10^7 hematologic cells with an efficiency of 96% is demonstrated.

Theory

The dielectrophoretic force is a frequency dependent function of a cell's volumetric and electrical properties, the applied electric field, and the suspending media.

$$\vec{F}_{DEP} = 2\pi\varepsilon_m r^3 Re\{f_{CM}\}\nabla(\vec{E}_{RMS} \cdot \vec{E}_{RMS}), \quad (1)$$

where ε_m is the permittivity of the suspending medium, r is the radius of the particle, and $\vec{E}_{RMS}$ is the root mean square electric field. $Re\{f_{CM}\}$ is the real part of the Clausius-Mossotti factor, defined as:

$$f_{CM} = \frac{\varepsilon_p^* - \varepsilon_m^*}{\varepsilon_p^* + 2\varepsilon_m^*}. \quad (2)$$

The complex permittivity is defined as $\varepsilon^* = \varepsilon - j\sigma/\omega$ where ε, and σ are the real permittivity and conductivity of the particle $_p$ or suspending medium $_m$, $j^2 = -1$, and ω is the radial frequency of the electric field. Due to their difference in size and electrical properties, cancer cells, white blood cells, and red blood cells experience different forces under the influence of the same electric field.

Materials and Methods

Microdevice

A schematic of the experimental cDEP device is shown in Figure 1. This device has a sample channel which is 5 mm wide, 10 mm long, and 50 μm deep. There are 4000 pillars which induce a large DEP force by generating non-uniformities in the electric field. The fluid electrodes are separated from the sample channel by a 20 μm barrier. The fluid electrodes were filled with highly conductive phosphate buffered saline (PBS). The geometry and size of these insulating pillars have been designed to optimize the fluid drag and DEP forces resulting in efficient trapping of cancer cells. The outlet was recorded for three minutes before and after turning on the electric field and number of cells passing through this region was counted. Three experiments were run for each cell type for each experimental parameter. Several control experiments were run without applying electric field and the number of cells entering the channel were counted and compared to the number of cells leaving the channel per unit of time. The percentage of trapped of cancer cells was then determined to be the ratio of the average number of cells passing through the outlet before and after the field was applied.

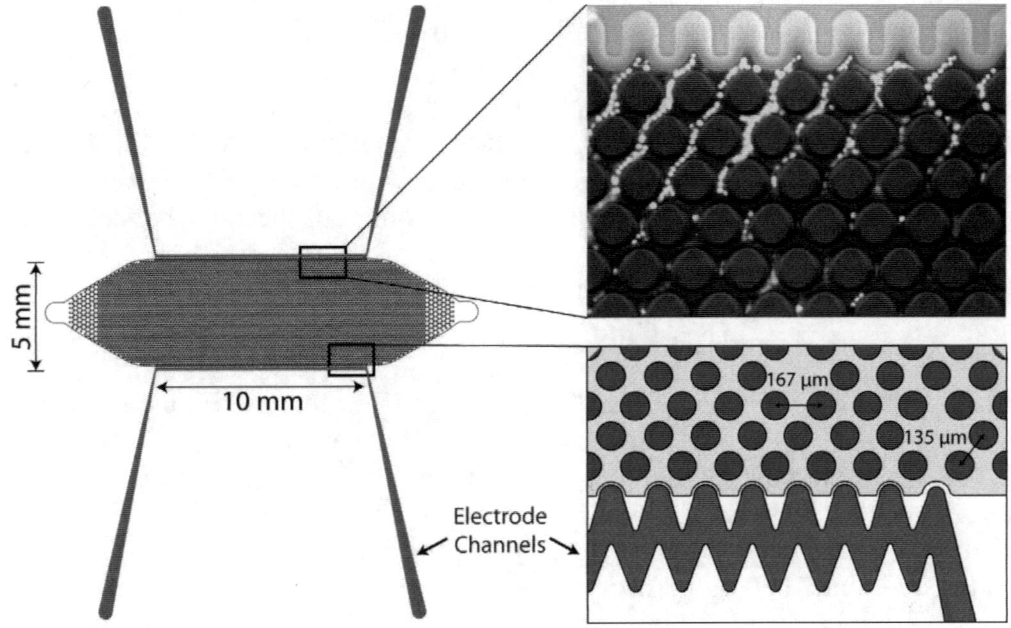

Figure 1. Top view schematic of the microdevice. MDA-MB-231 breast cancer cells are trapped when 300 V_{RMS} at 600 kHz is applied, while red blood cells pass freely in the background.

Blood Preparation:

Whole blood was extracted from a healthy donor into a vacutainer (BD Vacutainer ® Sodium Heparin, BD Franklin Lakes, NJ). The blood was centrifuged for 10 minutes at 3100 RPM (IEC Medilite, Thermo Scientific). The volume of supernatant was measured and removed from the sample via aspiration. The red blood cells were then washed three times in a low conductivity (DEP) buffer [14], to achieve a conductivity of 120 µS/cm as measured with a conductivity meter (Horiba B-173 Twin Conductivity/Salinity Pocket Testers, Cole-Parmer). The final volume of buffer added was equivalent to the volume of plasma removed. Sample preparation and experimentation were completed within 6 hours of extraction.

Cancer Cell Preparation

Samples of human three unique breast cancer cell lines (MDA-MB-231, MCF10A and MCF7), prostate cancer (PC3), and leukemia (THP-1) were individually washed twice and suspended in DEP buffer. Calcein AM, which is enzymatically converted to green fluorescent calcein, was added to cell sample at 2 μL/mL of cell suspension. The samples were then vortexed for 5 min, washed once and resuspended in DEP buffer. The cancer cells and blood sample were then mixed together in one conical tube with a final conductivity of 120 μS/cm. The fluorescently tagged cancer cells were easily identified among a large population of blood cells.

Computational Modeling

Computational modeling of the gradient of the electric field and velocity distributions can be used to effectively predict device performance and how cells will be manipulated within the device. The electric field distribution and its gradient, $\nabla E = \nabla(\nabla\phi)$, can be modeled computationally in Comsol multi-physics 4.2 using the AC/DC module (Comsol Inc., Burlington, MA, USA). This is done by solving for the potential distribution, ϕ, using the Laplace equation, $\nabla \cdot (\sigma^* \nabla\phi) = 0$. The boundary conditions used are prescribed uniform potentials at the inlet or outlet of the side channels. The electrical properties of PDMS used in the model are $\sigma_{PDMS} = 0.83 \times 10^{-12}$ S/m and $\varepsilon_{PDMS} = 2.65$. The electrical conductivity of PBS and DEP buffer are $\sigma_{PBS} = 1.4$ S/m and $\sigma_{DEP} = 0.01$ S/m, respectively, and their relative permittivities are $\varepsilon_{PBS} = \varepsilon_{DEP} = 80$.

Fluid dynamics related to the velocity field and shear rate in the main fluidic channel (Fig. 2(b)) were modeled using the COMSOL laminar flow module which solves the Navier-Stokes equations for incompressible flow. No slip boundary conditions were applied to the walls of channel and the pillars. The inlet velocity was set to 110 μm s^{-1} based on experimental volumetric flow rates and device geometry. The outlet was set to a zero viscous stress boundary condition. The viscosity and density of water were set as 0.001 Pa-s and 1000 kg m^{-3}, respectively.

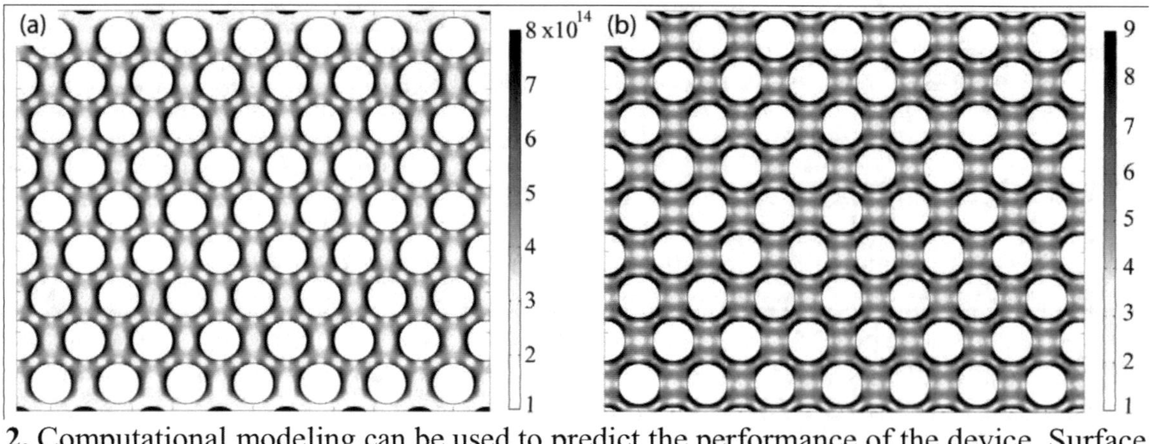

Figure 2. Computational modeling can be used to predict the performance of the device. Surface plot of (a) the gradient of the electric field squared (kg^2 m C^{-2} S^{-4}), and (b) shear rate (s^{-1}) in the fluidic channel.

Results and Discussion

Our computational results indicate that $\nabla\left|\vec{E}_{RMS} \cdot \vec{E}_{RMS}\right|$ is on the order of 10^{14} $kg^2mC^{-2}S^{-4}$ when 300 V_{RMS} was applied (Fig. 2(a)). This is approximately two orders of magnitude higher than the minimum threshold for cell manipulation in cDEP as reported previously. The maximum shear rate in the device, calculated to be 52 s^{-1}, was found to be below the threshold to cause cell lysis.

In the absence of an applied electric field, fouling of both cancer cells and blood cells was found to be negligible. When a high frequency AC electric field of sufficient voltage was applied, a cancer cells trapped near the pillars in the main channel (Figure 1). In contrast, red blood cells continued to flow through the device unaffected. This is likely due to the smaller radii of red blood cells. We it has previously been shown that the volumetric[15] and electrical[16] properties of these cells are different. Cell lysis was not observed during these experiments.

Figure 3 shows the percentage of trapped cancer cells as a function of applied voltage at 600 kHz for concentration of 1 cancer cell per 10^6 hematologic cells. We observed a similar trend for higher and lower concentrations of cancer cells. These results show that the percentage of trapping of cancer cells and efficiency of our technique are independent from cancer cell concentration. A maximum cancer cell trapping efficiency of 97% was achieved for THP-1 cells at 300 V_{rms} and 600 kHz.

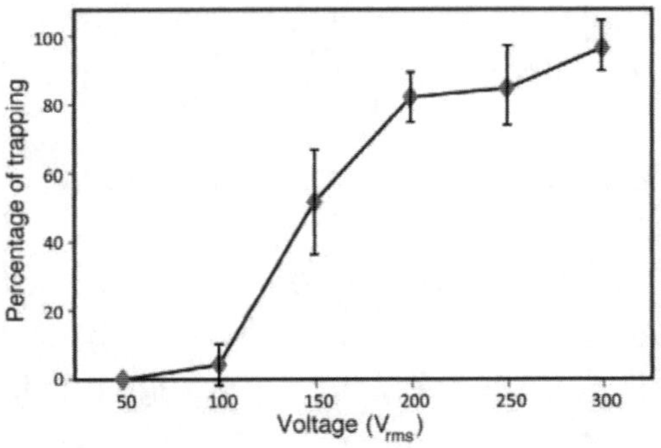

Figure 3. Percentage of trapping of cancer cells as a function of applied voltage at constant frequency of 600 kHz for concentration of 1 cancer cell per 106 hematologic cells.

Conclusions

In this study, a cDEP microdevice was presented which has a high selectivity towards cancer cells, a low selectivity towards blood cells, high flow rate, and low shear rate. The first two objectives were to ensure metastatic cells in low concentration were trapped while preventing clogging due to over enrichment of blood cells. A high flow rate is necessary to seek out the hypothesized a few CTCs in 1 mL of blood. Finally, a low shear rate is essential to prevent cell damage due to the velocity of the sample fluid. Using cDEP technique we succeed to isolate prostate, leukemia, and three different stages of breast cancer cell lines from blood while the concentration of cancer cells was as low as 100 cells per 1 mL of blood.

To our knowledge, this study is the first study using non-diluted blood in DEP. In our method of blood preparation, we keep all of the complexity of blood including all of the blood cells. This method in also much easier than some other methods of blood sample preparation for DEP experiments[15]. Based on these results, we hypothesize that these cDEP devices can be used for clinical isolation and detection of CTCs for early cancer detection.

Acknowledgment

This material is based upon work supported in part by the National Science Foundation under Grant No. EFRI 1107049 and by the Virginia Tech Institute for Critical Technology and Applied Science (ICTAS).

References

1. M. Yu, S. Stott, M. Toner, S. Maheswaran and D. A. Haber, *The Journal of cell biology*, 2011, **192**, 373.
2. H. Shafiee, J. L. Caldwell, M. B. Sano and R. V. Davalos, *Biomed Microdevices*, 2009.
3. H. A. Pohl, *Dielectrophoresis: The behavior of neutral matter in nonuniform electric fields*, Cambridge University Press, Cambridge, 1978.
4. H. Shafiee, M. B. Sano, E. A. Henslee, J. L. Caldwell and R. V. Davalos, *Lab Chip*, 2010, **10**, 438-445.
5. Y. Huang, S. Joo, M. Duhon, M. Heller, B. Wallace and X. Xu, *Anal Chem*, 2002, **74**, 3362-3371.
6. J. Cheng, E. Sheldon, L. Wu, A. Uribe, L. Gerrue, J. Carrino, M. Heller, and J. O'Connell, *Nat Biotechnol*, 1998, **16**, 541-546.
7. L. Altomare, M. Borgatti, G. Medoro, N. Manaresi, M. Tartagni, R. Guerrieri and R. Gambari, *Biotechnol Bioeng*, 2003, **82**, 474-479.
8. G. H. Markx, M. S. Talary and R. Pethig, *Journal of Biotechnology*, 1994, **32**, 29-37.
9. Y. Huang, X. B. Wang, F. F. Becker and P. R. C. Gascoyne, *Biophys J*, 1997, **73**, 1118-1129.
10. U. Kim, C. W. Shu, K. Y. Dane, P. S. Daugherty, J. Y. J. Wang and H. T. Soh, *P Natl Acad Sci USA*, 2007, **104**, 20708-20712.
11. A. Lostumbo, D. Mehta, S. Setty and R. Nunez, *Exp Mol Pathol*, 2006, **80**, 46-53.
12. S. Nagrath, L. V. Sequist, S. Maheswaran, D. W. Bell, D. Irimia, L. Ulkus, M. R. Smith, E. L. Kwak, S. Digumarthy, A. Muzikansky, P. Ryan, U. J. Balis, R. G. Tompkins, D. A. Haber and M. Toner, *Nature*, 2007, **450**, 1235-1239.
13. K. Kato and A. Radbruch, *Cytometry*, 1993, **14**, 384-392.
14. L. A. Flanagan, J. Lu, L. Wang, S. A. Marchenko, N. L. Jeon, A. P. Lee and E. S. Monuki, *Stem Cells*, 2008, **26**, 656-665.
15. H.-S. Moon, K. Kwon, S.-I. Kim, H. Han, J. Sohn, S. Leeb and H.-I. Jung, *Lab Chip*, 2011, **11**, 1118–1125.
16. P. R. C. Gascoyne, X. B. Wang, Y. Huang and F. F. Becker, *Ieee T Ind Appl*, 1997, **33**, 670-678.

(526b) Design, Development, and Analysis of a Multilayer Contactless Dielectrophoresis Device to Separate Cancer Cells From Blood

Michael B. Sano, School of Biomedical Engineering & Sciences, Virginia Polytechnic Institute and State University, Blacksburg, VA and Rafael Davalos, School of Biomedical Engineering and Sciences, Virginia Polytechnic Institute and State University, Blacksburg, VA

The use of non-invasive methods to detect and enrich cancer cells independent of their genotype is critical for early diagnostic and treatment purposes. In addition, isolation of these cells could provide a workbench for clinicians to screen drug therapies prior to patient treatment. This would enable oncologists to tailor treatment on a patient specific level and to ensure the most effective treatment is being utilized. Since many symptoms associated with metastatic cancers, can be attributed to multiple diseases, diagnosis of this disease must be accomplished via medical analysis. There are a number of molecular, cytogenic, immunological cytochemical and morphological assays capable of making this diagnosis, however, the small ratio of cancerous to normal cells make these processes challenging.

During progression, tumor cells acquire the capacity to disseminate into blood circulation until they either are detected and eliminated by the immune system or attach to the endothelial cells, extravasate, and grow as secondary tumors (metastasis) at distant sites. These circulating tumor cells (CTCs) could provide early indicators of metastasis. Typically, techniques for detecting these cells require prior knowledge of cell-specific markers and antibodies. The large amount of sample handling required by these techniques may increase the likeliness of cell loss and contamination. Therefore, there is a growing need for marker-independent isolation and purification method to increase yield, sensitivity, precision and reproducibility.

Dielectrophoresis (DEP), the motion of a particle in a non-uniform electric field, has become a robust method for analyzing nano-particles, cells, viruses, and DNA based on their physical and electrical properties. A new technique, contactless Dielectrophoresis (cDEP), isolates cells from contact with the electrodes. This is achieved by using fluid electrodes which are isolated from the sample channel by thin insulating membranes. The absence of contact between electrodes and the sample fluid inside the channel prevents bubble formation and avoids any contaminating effects the electrodes may have on the sample. This technique has demonstrated the ability to isolate THP-1 human leukemia monocytes from a heterogeneous mixture of live and dead cells as well as discriminate between cells of different metastatic potential. Recently, this technique has been improved for operation at frequencies as low as 1 kHz, allowing us to manipulate cells using both positive and negative DEP. This work presents the development, analysis, and evaluation of a three dimensional, multilayer, cDEP device used to separate cancer cells from blood.

(526c) Sterilization of Bacteria by Pulsed Electric Field On Microfluidic Chip

Jun Yang[1], **Ning Hu**[1] and Shizhi Qian[2], (1)Bioengineering College, Chongqing University, Chongqing, China, (2)Department of Mechanical and Aerospace Engineering, Old Dominion University, Norfolk, VA

High temperature heat sterilization technique has been widely used for the sterilization of water and foods due to its simplicity and high efficiency. However, this process is energy-intensive and the high temperature can adversely affect nutritional qualities and flavor of food. Some of the aforementioned shortcomings can be overcome by the pulsed electric field (PEF) sterilization technique, which is nonthermal and uses short-duration (tens of ns to 100 ms) high-field electric pulses to sterilize liquid foods such as fruit juices, beer, milk, and cream. However, due to large distance between the two counter electrodes used in the typical PEF systems, high-voltage pulses provided by an expensive high-voltage pulse generator are required to achieve high electric field. With the advancement of microfabrication technologies, high electric field pulses can be easily generated by a low-voltage pulse generator when the two counter electrodes are a few hundred micron meters apart. A microfluidic chip for sterilization of bacteria using the PEF sterilization technique has been designed and fabricated. The side walls of the microfluidic channel also serve as the two counter electrodes, which are 100 mm apart. The performance of the chip was tested by sterilization of *Escherichia coli* under various conditions.

See more of this Session: Electroporation, Electrophysiology and Cell Electrokinetics
See more of this Group/Topical: Topical 3: 2011 Annual Meeting of the American Electrophoresis Society (AES)

(526d) Nanoelectroporation and Its Comparison with Micro and Bulk Electroporation

Wei-Ching Liao, Mechanical and Aerospace Engineering, The Ohio State University, Columbus, OH, Pouyan E. Boukany, Nanoscale Science and Engineering Center for Affordable Nanoengineering of Polymeric Biomedical Devices, The Ohio State University, Columbus,, OH, Hyun Chul Jung, Electrical and Computer Engineering, The Ohio State University, Columbus, OH, Wu Lu, Department of Electrical & Computer Engineering, The Ohio State University, Columbus, OH, Gregory Lafyatis, Department of Physics, The Ohio State University, Columbus, OH and L. James Lee, Chemical and Biomolecular, the Ohio State University, Columbus

A single cell nanoelectroporation (NEP) technology has been developed in our research group. In our NEP process, a hybrid micro/nanochannel device fabricated by a DNA combing and imprinting (DCI) technique was used to electroporate individual cells at the tip of nanochannel. Injective/convective material transport was found in the NEP to provide the capability of giving cell a shot with minimum cell damage, which is not achievable by the diffusion-dominated microelectroporation (MEP) and bulk electroporation (BEP) processes because of different delivery mechanisms. In this study, systematic comparison among NEP, MEP and BEP is investigated based on both theoretical and experimental analyses to elucidate the unique characteristics of NEP. The mechanism differences among NEP, MEP, and BEP are investigated through the delivery of PI dye and quantum dots (QD) in the experiments with electric current measurements. Numerical simulation is carried out to study the detailed electric field and material electrophoresis dynamics in each electroporation process. From the observed transport profiles and computed transmembrane potential (TMP) distributions and electrophoretic and diffusive transports, NEP is proved to be a unique single cell electroporation tool for does-controlled transfection.

See more of this Session: Electroporation, Electrophysiology and Cell Electrokinetics
See more of this Group/Topical: Topical 3: 2011 Annual Meeting of the American Electrophoresis Society (AES)

(526e) A 3D Bulk Microelectrode Array for High Throughput Cell Electrofusion

Ning Hu[1], Sang Woo Joo[1], Jun Yang[2], Shizhi Qian[3] and Xiaolin Zheng[2], (1)School of Mechanical Engineering, Yeungnam University, Gyongsan, South Korea, (2)Chongqing University, Chongqing, China, (3)Department of Mechanical and Aerospace Engineering, Old Dominion University, Norfolk, VA

Cell electrofusion is a state-of-the-art technology that applies electric field to induce two or more cells to merge into a hybrid cell in an asexual way. Since the hybrid combines genetic and epigenetic information of two cells, it can be used for gene expression analysis, reprogramming somatic cells, developing antibody, cloning mammals, and cancer immunotherapy, among other important applications. In this work, a micirofluidic chip which integrates a novel 3D bulk microelectrode array is used for cell electrofusion under a low applied voltage. On this chip, insulating bulk is used to fill the space between two adjacent protruding microelectrodes on the same side of the microchannel. With this design, cell trapping in undesirable spaces reported previously can be avoided. The 3D bulk microelectrode array is fabricated on a silicon-on-insulator (SOI) wafer by lithography technology with 4 masks. The 35 mm thick highly doped Si layer is used for the microelectrode array and the wall of the microchannel. Insulating gap with the same depth is fabricated between the two adjacent protruding microelectrodes. A smooth flow thus is formed, and the protruding microelectrode structures is used to generate non-uniform electric field distribution. The gaps between any two adjacent protruding microelectrodes on the same side of the microchannel are filled by polycrystalline silicon after high temperature oxidation of the surface of the side walls. The highly doped Si layer is covered by a thin aluminum film to improve the conductivity. A SiO_2 film is subsequently deposited on the top of these microelectrodes for better biocompatibility, resistance to oxidation, and corrosion. In cell alignment and pairing research, almost all cells (>99%) were driven by the dielectrophoretic (DEP) force to the high electric field strength area to be aligned as cell-cell pairs. After the alignment, low voltage pulses were loaded on the microelectrode to generate high strength electric filed to induce reversible electroporation of the cytomembrane. Reconstructing cytomembrane, exchanging cytoplasm and forming hybrid can be completed within 3 min. Due to the non-uniform electric-field distribution, two-cell fusion was ensured in this device, even when more than two cells are formed as a multi-cell chain. On this chip, higher fusion efficiency than traditional chemical method (less than 5%) and electrofusion methods (less than 12%) has been achieved.

See more of this Session: Electroporation, Electrophysiology and Cell Electrokinetics
See more of this Group/Topical: Topical 3: 2011 Annual Meeting of the American Electrophoresis Society (AES)

(526f) Developing a Method for the Isolation of Intact Viable Giant Mitochondria Utilizing An Electric Field

Scott Rose and Vratislav Kostal, Chemistry, University of Minnesota, Minneapolis, MN

Developing a Method for the Isolation of Intact Viable Giant Mitochondria Utilizing an Electric Field

Scott P. Rose, Vratislav Kostal and Edgar A. Arriaga (Department of Chemistry, University of Minnesota-Twin Cities)

Introduction and Background: Observable changes in mitochondrial morphology occur in shape and size as the organism ages. Giant mitochondria can range between 1-5 μm, but the ability to analyze this unique organelle has been limited due to their fragile nature. Current methods of organelle isolation deal with mechanical homogenization, which creates a powerful shear force that work well to release sub-cellular components, but tear apart the giant mitochondria and subsequently make analysis not possible. We propose that the use of an electric field, applied with an electroporator, in conjunction with a solution containing digitonin, trypsin and colchicine will allow us to safely remove the plasma membrane and cytoskeleton from the mitochondria, thereby keeping the organelles of interest viable and available for study.

Method: 10-*N*-nonyl acridine orange NAO labeled L6 rat myoblasts are being studied in a mitochondrial-isolation buffer solution in combination with trypsin, colchicine and digitonin. Sample solution is then exposed to 90, 0.35kV pulses from an electroporator. After electroporation, cells are then observed using fluorescent microscopy and flow cytometry or Capillary Electrophoresis with Laser Induced Fluorescence (CE-LIF) to determine cell count and number of intact mitochondria.

Preliminary Work: Our current method, using 90 short, electrical chain pulses disrupts upwards of 90% of the plasma membranes. The presence of mitochondria in an isolated subcellular fraction can be verified by flow cytometry and Capillary Electrophoresis with Laser Induced Fluorescence. Further analysis is being conducted using Flow Cytometry to identify the presence of giant mitochondria separate from whole cells and regular sized mitochondria.

See more of this Session: Electroporation, Electrophysiology and Cell Electrokinetics
See more of this Group/Topical: Topical 3: 2011 Annual Meeting of the American Electrophoresis Society (AES)

(526g) Capillary Electrophoresis of Magnetically Enriched Autophagosomes

Chad Satori[1], Vratislav Kostal[1] and Edgar A. Arriaga[2], (1)Chemistry, University of Minnesota, Minneapolis, MN, (2)Department of Chemistry, University of Minnesota, Minneapolis, MN

Autophagy is one of the main cellular degradation processes. This process involves autophagosomes that are organelles credited with accumulating subcellular components marked for degradation. These organelles then fuse with lysosomes to complete degradation. It has been suggested that the lipid composition of autophagosomes play an important role in autophagy, fusion with other organelles, and signaling pathways but these ideas have not been thoroughly explored. It is also unknown how changes in lipid content or proteome impact autophagosome function in lysosome storage and neurodegenerative disorders. The ability to separate and characterize an enriched autophagosome fraction will increase the ability to study these research needs of autophagosomes.

This presentation will report on recent advances to analyze individual autophagosomes by capillary electrophoresis with laser-induced fluorescence detection (CE-LIF). Whole cell homogenate from EGFP-LC3 (enhanced green fluorescent protein-LC3 protein) transfected cells is prepared and treated with anti-LC3 labeled magnetic microparticles. Microparticles and bound organelles are separated from the rest of the biological contents with a custom-built on-line magnetic separator. This fraction is then analyzed for EGFP fluorescence using CE-LIF.

See more of this Session: Electroporation, Electrophysiology and Cell Electrokinetics
See more of this Group/Topical: Topical 3: 2011 Annual Meeting of the American Electrophoresis Society (AES)

() High Sensitivity Insulator Based Dielectrophoretic Phenotyping of Bacteria

William Braff[1], Dana L. Willner[2], Phil Hugenholtz[3], Korneel Rabaey[2] and **Cullen R. Buie**[1], (1) Department of Mechanical Engineering, Massachusetts Institute of Technology, Cambridge, MA, (2) University of Queensland, St. Lucia, Australia, (3)Australian Center for Ecogenomics, University of Queensland, St. Lucia, Australia

We have recently designed a novel platform for screening cells known as three-dimensional insulator based dielectrophoresis (3DiDEP). In dielectrophoresis (DEP), microorganisms in a non-uniform electric field experience forces related to their volume and dielectric properties. Differences in physical properties can be exploited to separate populations of cells. To date DEP has been used to sort cells based on morphology (e.g. gram positive vs. gram negative bacteria), size, and viability. 3DiDEP features advantages over previous embodiments of DEP in that it offers high sensitivity, uses low applied voltages, requires minimal fabrication, and is relatively simple to operate. Our recent work has shown that 3DiDEP can be useful to distinguish bacteria with sub-species resolution, in some cases better resolution than 16S rRNA sequencing. In this talk I will discuss our 3DiDEP design and describe some exciting preliminary results on the characterization of pathogenic bacteria.

See more of this Session: Electroporation, Electrophysiology and Cell Electrokinetics
See more of this Group/Topical: Topical 3: 2011 Annual Meeting of the American Electrophoresis Society (AES)

(565a) History First Award Session of the American Electrophoresis Society

Lawrence I. Grossman, Center for Molecular Medicine and Genetics, Wayne State University, Detroit, MI and **David E. Garfin**, American Electrophoresis Society, Madison, WI

A brief history about the American Electrophoresis Society will be presented, including important milestones such as the 10th anniversary of holding the AES annual meeting in conjunction with the AIChE annual meeting.

An introduction to the first Award Session of the American Electrophoresis Society in Honor of Dr. Kelvin Lee will be presented as well.

See more of this Session: Award Session of the American Electrophoresis Society In Honor of Kelvin Lee
See more of this Group/Topical: Topical 3: 2011 Annual Meeting of the American Electrophoresis Society (AES)

(565b) The Importance of Electrophoresis In Protein Analysis: From Sandefjord and Wild Dunes to Minneapolis

Kelvin H. Lee, Department of Chemical Engineering, University of Delaware, Newark, DE

The development of technologies to facilitate the electrophoretic separation of proteins can rely on fundamental principles of chemical engineering. In addition, there are a wide variety of applications of protein electrophoresis which can begin to address important problems in bioprocess engineering and biomedical engineering. This presentation will review a series of research efforts that begin with protein analysis technologies and include applications of two-dimensional protein electrophoresis in the diagnosis and understanding of neurodegenerative disease. An overarching theme is the impact of American Electrophoresis Society members and meetings on these efforts.

See more of this Session: Award Session of the American Electrophoresis Society In Honor of Kelvin Lee
See more of this Group/Topical: Topical 3: 2011 Annual Meeting of the American Electrophoresis Society (AES)

(565c) Optimization of Sample Preparation for Two-Dimensional Protein Electrophoresis

Kristin N. Valente, Leila H. Choe, Abraham M. Lenhoff and Kelvin H. Lee, Department of Chemical Engineering, University of Delaware, Newark, DE

Effective sample preparation is essential to capture the sample proteome and maximize utility of subsequent electrophoresis. The large number of interacting variables impacting protein solubility and the relatively low-throughput nature of most electrophoretic methods constrains the systematic study of sample preparation. We employed a design of experiments approach to optimize primary protein recovery by altering solution compositions during sample preparation. The impact of four components (urea, DTT, CHAPS, and SDS) was examined, with protein recovery quantified by protein concentration assays, SDS-PAGE, and two-dimensional electrophoresis. Recovery results were obtained for a mixture of model proteins, as well as CHO-K1 and *E. coli* cell lysates, and favorable solution conditions for each sample type are discussed.

See more of this Session: Award Session of the American Electrophoresis Society In Honor of Kelvin Lee
See more of this Group/Topical: Topical 3: 2011 Annual Meeting of the American Electrophoresis Society (AES)

(565d) Control In Microfluidic Devices

Mark A. Burns, Chemical Engineering, University of Michigan, Ann Arbor, MI

Microfluidic devices have the potential to have a broad impact on society through the miniaturization and massive parallelization of chemical assays and experiments. Many separation, reaction, and transport components have been developed for these devices but the major issues that arise when designing complex system can often be traced to the control of the components. Control of some components is relatively straightforward: the temperature in a reactor can be controlled by simply fabricating metal lines (i.e., heaters and temperature sensors) near the chamber. Other situations, such as migrating bands in an electrophoresis gel or the position of fluid samples in channels, require more complex solutions. In this talk, I will describe several examples of both problems and solutions of these control issues.

See more of this Session: Award Session of the American Electrophoresis Society In Honor of Kelvin Lee
See more of this Group/Topical: Topical 3: 2011 Annual Meeting of the American Electrophoresis Society (AES)

(565e) Effective Separations: The Key for Proteomics

Phillip C. Wright, ChELSI Institute, Department of Chemical and Biological Engineering, The University of Sheffield, Sheffield, United Kingdom

Proteomics seeks to study the total proteins expressed in any given system, whether by abundance, activity, structure, state of post translational or other modification or how these proteins interact with each other in networks or complexes. A key component to the proteomics workflow is separation. Crude proteomes are the consequence of the expression of thousands of genes, with complexities made even greater by the presence of PTMs and large variation in dynamic ranges (potentially exceeding 9 orders of magnitude). To address these dynamic range issues, proteomes need to be simplified in order to enable efficient analysis. The proteomics practitioner has a number of choices to make when seeking to examine a proteome and this then may dictate the 'simplification' process one chooses when performing a deeper analysis for systems biology endeavors. Traditionally, the choice was based on one-dimensional or two-dimensional gel electrophoresis, but this has now extended out to include various 'flavors' of chromatography and gas-phase fractionation in the mass spectrometer. This study seeks to demonstrate the advantages of selected (pre)separation strategies on identifying and quantifying proteomes from the three domains of life. The modern proteomics experiment relies heavily on the mass spectrometer (MS) and care must be taken to understand the relationship between the separation technology and the MS. Examples of post translation modification are given, as well as those that to seek to quantify the proteome using label-based methods such as iTRAQ. The impact of the separation strategy on the quality and quantity of data are provided.

See more of this Session: Award Session of the American Electrophoresis Society In Honor of Kelvin Lee
See more of this Group/Topical: Topical 3: 2011 Annual Meeting of the American Electrophoresis Society (AES)

AUTHOR INDEX

AUTHOR INDEX

AUTHOR INDEX